Commonsense Guide to Current Affairs

Commonsense Guide to Current Affairs

The Issues we Read and Hear about Every Day from the Standpoint of what the Politicians have Forgotten—Common Sense

Vincent Frank Bedogne
and
Marcy Jean Everest

RESOURCE *Publications* • Eugene, Oregon

COMMONSENSE GUIDE TO CURRENT AFFAIRS
The Issues we Read and Hear about Every Day from the Standpoint of what the Politicians have Forgotten—Common Sense

Copyright © 2009 Vincent Frank Bedogne. All rights reserved. Except for brief quotations in critical publications or reviews, no part of this book may be reproduced in any manner without prior written permission from the publisher. Write: Permissions, Wipf and Stock Publishers, 199 W. 8th Ave., Suite 3, Eugene, OR 97401.

Resource Publications
A Division of Wipf and Stock Publishers
199 W. 8th Ave., Suite 3
Eugene, OR 97401
www.wipfandstock.com

ISBN 13: 978-1-60608-786-2

Manufactured in the U.S.A.

To Bob Everest

. . . In a time when we sue Burger King for serving burgers and picket Kentucky Fried Chicken for serving chicken, we forget the obvious. In an age when we navigate the rapids of political correctness in a headlong plunge to this or that job, to this or that school, to this or that purchase, to this or that wife, husband, boyfriend, or girlfriend, one element is missing. It's not ability or imagination. It's not drive, passion, or ambition. It's not economic ideology, political ideology, or religious ideology. It's common sense.

Contents

Acknowledgments ix
Preface xi

1 Argument 1
2 Science 5
3 Statistics 9
4 Diet 13
5 Exercise 18
6 Self 21
7 Collective 25
8 Family 29
9 Society 34
10 Civilization I 38
11 Scarcity 42
12 Civilization II 46
13 Economics 51
14 Civilization III 56
15 Socialism 61
16 Civilization IV 65
17 Capitalism 70
18 Civilization V 74
19 Today's Economy 78
20 World War I 85

21	Taxes 90
22	World War II 95
23	Politics 100
24	Cold War 103
25	Democracy 109
26	Education 114
27	Healthcare 120
28	Earth 126
29	Environment 131
30	Population 136
31	Global Warming 141
32	Cap and Trade 146
33	Energy 153
34	Hybrid Cars 158
35	Renewable Energy 163
36	Hydrogen 169
37	Nuclear Power 173
38	Energy Plan for America 179
39	The Middle East 185
40	God 191
41	Islam 195
42	Terrorism 200
43	Darwin 206
44	Clones 211
45	Abortion 214
46	UFOs 220
	Bibliography 225

Acknowledgments

The authors would like to express their appreciation to their families and to all whom contributed to the book and offered their comments and criticisms. Particular gratitude goes to those who had the greatest influence on their thoughts and who helped with the book's editing and manuscript preparation. Above all, Tom Pyle, Jim Tedrick, Diane Farley, Henry Bettich, Raydeen Cuffe, Andrew Bedogne, Bill Meulemans, Heather Carraher, Christian Amondson, and Tina Campbell Owens.

Preface

THE ONE THING WE can say for sure about today's world is that it's crazy.

Liberals hate conservatives. Democrats hate Republicans, and the last straight answer we've gotten out of a president was in the Roosevelt administration, Teddy. Prices go up and down, but more up than down. Wages go up and down, but more down than up—and when they do go up they never go up enough to cover increasing prices. One politician says that the economy is structurally sound and poised for a quick recovery, another that it's in the worst shape since the Great Depression and we should prepare for years of pain and struggle.

Shiite Muslims are often at odds with and don't see eye-to-eye with Sunnite Muslims. Sunnite Muslims are often at odds with and don't see eye-to-eye with Kurdish Muslims. Shiite, Sunnite, and Kurdish Muslims are often at odds with and don't see eye-to-eye with Wahhabi Muslims. And they all are often at odds with and don't see eye-to-eye with Americans, who can't figure out, and for the most part don't care, if Colin Powell is black or white or if Barack Obama is black enough much less the relationship between Shiite, Sunnite, Kurdish, and Wahhabi Muslims. Where again is the Middle East?

Hot off the press, the latest, sexiest rap star gets shot. The latest, sexiest sports star gets arrested. The latest, sexiest pop star gets divorced. The latest, sexiest film star enters rehab. Download today's celebrity mug shot, the nude version please, but use a proxy server to access an encrypted peer-to-peer so that the music industry won't think you're stealing a song, seize your computer, and sue you for a million dollars.

Eat eggs. Don't eat eggs. Eat meat. Don't eat meat. Eat pasta. Don't eat pasta. Eat vegetables. Don't eat vegetables. Fat is bad. Fat is good. It's all in your mind. A new diet comes out every week, and the latest nutrition study concludes that what the previous study concluded would doom us to a statistically significant risk of death we can't live without.

Parents refuse to immunize their kids, which won't matter if the environmentalists are right and there will be no fish in the oceans by 2049 and the earth will be dead by 2050. Which won't matter if the New Agers—who are getting old—are right and the UFOs will whisk us back in time, forward in time, to another dimension, or to a lost civilization on Mars.

And it's all TV's fault.

Our world is crazy, out-of-control, rushing in every possible direction. The only thing we can count on is that Oprah will be around to, between celebrity makeovers, tell us to love ourselves, and Jerry Springer will be around to—with "YouTube" memories of fistfights, cat fights, midget fights, Steve the security guy, and Doctor Butterworth the psychologist—ask why we can't all just get along. In a time when we sue Burger King for serving burgers and picket Kentucky Fried Chicken for serving chicken, we forget the obvious. In an age when we navigate the rapids of political correctness in a headlong plunge to this or that job, to this or that school, to this or that purchase, to this or that wife, husband, boyfriend, or girlfriend, one element is missing. It's not ability or imagination. It's not drive, passion, or ambition. It's not economic ideology, political ideology, or religious ideology. It's *common sense*.

In the chapters ahead, we look at the topics and issues that affect our lives from the standpoint of what we know to be sound, our all-too-often repressed ability to see things in a practical and matter-of-fact way. Argument, Science, and Statistics. Diet, Exercise, Self, Collective, Family, and Society. Socialism, Capitalism, Taxes, Politics, Democracy, Education, Healthcare, and the ascent of civilization. Earth, Environment, Population, Energy, Global Warming, Hybrid Cars, Hydrogen, and Nuclear Power. God, Islam, Terrorism, Darwin, Clones, Abortion, and UFOs. All in the light of our ability to filter what we read and watch though the sieve of common sense.

And, as we learn to better use our common sense, we discover something remarkable. The complex and unsolvable problems of our time don't seem quite as complex or quite as unsolvable. Common sense—that which we have gained through our day-to-day trials and have had within us all along—empowers us. It gives us a strength to tear through the fabric of facts, figures, and anecdote draped over our minds by the powers that attempt to mold our thoughts. Reporters no longer seem aloof. Pundits no longer seem all-knowing. Politicians no longer seem in command of it all. And beyond the media blind—and the careers, agendas, factions, ideologies, and personalities it represents—do we find the truth? Perhaps. At the least, we grow strong in our quest for the truth and firm in our conviction to settle for nothing less.

1

Argument

To apply our common sense to the issues of today, we need certain skills. First off is the technique that writers and debaters call the *argument*. By argument, we don't mean what goes on when we're angry with our kids, spouse, or parents. Argument isn't a shouting match or a fistfight. Formally, it's the technique we use and that is used by others countless times every day to make a point. The better we understand how an argument works, the better we can get our points across and determine the soundness of those made by others.

Technically, argument falls into the area of philosophy called *logic*, and every argument has a specific form. Arguments consist of statements, of which one or more, the *premises*, provide support for, or reasons to believe the others, the *conclusions*.

Premise: This quarter's leading economic indicators are up.
Conclusion: Therefore, the economy is doing better.

How many times have we heard this argument or its converse: This quarter's leading economic indicators are down, therefore the economy isn't doing as well. In either case, a premise or statement of fact, the economic indicators, supports a conclusion, the economy is getting better or worse.

Arguments always have at least one premise and one conclusion, but they aren't always as easy to pick out of an article or discussion as our economics example might suggest. Consider the following example, derived from a 2003 National Institute of Allergy and Infectious Diseases report[1]

---

1. "Many Americans Think an AIDS/HIV Vaccine Already Exists."

with regard to AIDS, or Acquired Immune Deficiency Syndrome, and the Human Immunodeficiency Virus, HIV, believed to cause it:

> A government and pharmaceutical industry cover-up prevents the use of an HIV vaccine. A recent study says that 48 percent of Hispanics, 28 percent of African Americans, and 20 percent of the general population believe an HIV vaccine exists.

In this example, the argument is implied by the overall statement.

Premises: An HIV vaccine exists and is not in use.

Conclusion: Therefore, the government and pharmaceutical industry have conspired to cover it up.

Premises and conclusions rarely stand out but are entwined with facts, examples, and background. This may be done to mask an unsound argument but most often is done to provide context and to make the prose easier to read. As writers, we want the words to flow.

As by now you may have figured, there are two categories of arguments: *good arguments* and *bad arguments*. A good argument is one in which the premises support the conclusions. A bad argument is one in which the premises claim to support the conclusions but fail to do so. Our HIV and economics examples illustrate good arguments. You may disagree with their conclusions, but as stated they logically follow from the premises. In contrast:

> Global meteorological monitoring the last fifty years indicates that the earth's average temperature has increased. Therefore, sport utility vehicles should be banned.

In this argument—one that if you drive a sport utility vehicle, or SUV, you hear more than perhaps you'd like—the premise, the earth's temperature has increased, doesn't support the conclusion, SUVs should be banned. There is no direct or stated relationship between the two—unless, as you may have unconsciously done, one fills in the blanks with certain preconceptions about global warming.

A bad argument is always bad. A good argument, logical though it may be, however, need not be a *sound* argument, one that gets the point across in a way we accept or feel confident making. For an argument to be sound, the premises must be true, or such that we can reasonably assume they're true, and the argument must embody a clear line of reasoning that leads to the conclusion.

Derived from an opinion survey, our HIV argument is predicated on the statement that a vaccine for the disease exists. But fact and belief aren't necessarily the same. Is the premise self-evident? Is it supported by research? Can we be confident of its validity? Of course not. Our HIV example is a "good" but not a "sound" argument. An argument based on a false premise may be valid from the standpoint of logic but will support a false conclusion.

Similarly, as our SUV example illustrates, a sound argument must embody a reasoning process that leads from the premises to the conclusions. What if we rewrite our SUV argument?

> **Premises: Global meteorological monitoring the last fifty years indicates that the earth's average temperature has increased. Global warming adversely affects climate. In the lab, carbon dioxide creates a greenhouse effect, which increases temperature. Sport utility vehicles emit carbon dioxide.**
>
> **Conclusions: Sport utility vehicles increase global warming and should be banned.**

Stated as such, a coherent line of logic leads from the premises to the conclusions. We have turned a bad, logically incorrect, argument into a good argument and, at least from the standpoint of a reasoning process, into a sound argument.

Taken as whole, however, our SUV argument is still weak. Carbon dioxide does in some instances, specifically with regard to certain wavelengths of light, create a greenhouse effect in the lab, and SUVs do emit carbon dioxide. These are observed phenomena, facts. The earth's temperature change, the detrimental climatic effects of global warming, the extent to which natural as opposed to manmade variables come into play, and the behavior of carbon dioxide in the atmosphere as opposed to the lab are open for discussion, not facts.[2]

We can further refine our understanding of the argument by incorporating two additional concepts: *deductive* and *inductive* reasoning. A deductive argument is one in which, if the premises are assumed to be true, the conclusion must be true. An inductive argument is one in which,

---

2. See chapter 31, *Global Warming*.

if the premises are assumed to be true, it's probable but not certain that the conclusion is true.

Our economics example illustrates a deductive argument. Because we measure economic strength with indicators, if they go up or down, the economy by definition gets better or worse. This is not to say that economic indicators such as inflation and unemployment are necessarily valid. Looked at critically, their accuracy is debatable. But if we take them to be a measure of economic activity, the conclusion definitively follows the premise.

Though unsound, our HIV example illustrates an inductive argument. If we take the premise to be true and assume a vaccine for the diseases exists, we infer, rather than deduce, that since our doctor hasn't heard about it certain factions are conspiring against its use. Paranoia and late-night talk-radio aside, this is a possible but not an absolute conclusion. The vaccine could be unavailable because it's hard to produce or has yet to be proven safe and effective.

Every day we make arguments and we face arguments. An argument has premises and conclusions. If the premises support the conclusions it's a good argument, if not it's a bad argument. Though logical, a good argument isn't necessarily a sound argument, one that gets the point across in a reasonable way. To be sound, the premises must be true, or such that we can confidently assume they're true, and the argument must embody a clear line of reasoning that leads from the premises to the conclusions. Moreover, an argument may allow us to deduce or to infer a conclusion. Empowered with the skill of the argument, we know how to get our points across to others, and what we read and hear loses its sense of authority. We have the ability to probe beyond the words, to critically examine the ideas spewed at us every day. And we will do so throughout this book.

2

Science

To make or interpret a point, we apply the skill of the argument. A sound argument rests on a premise that states a fact or something we know or with a high degree of certainty assume to be true. But how do we determine if a premise is actually true? Sometimes it's self-evident, something we know from experience or common sense. Fire is hot. Ice is cold. The fastest man on earth can't outrun a bus. Sometimes it's based on definition. By definition, infinity is infinite. By definition, emptiness is empty. By definition, division by zero is undefined. Sometimes the truth of a premise is based on faith, on conviction. There is a God. There is an afterlife. There is hope for a better future. Most of the time, it's based on science.

In modern society, science occupies a special place in our lives. What technological advance hasn't come from science? What poll or survey doesn't claim to be scientific? What drug or nutritional supplement doesn't claim to be clinically tested? We invoke the name of science to argue almost every point: technical, political, ideological. Science is at the core of our existence, pivotal to the function of our world. But what is science?

If we think back to our grade school days, we recall that science has something to do with a hypothesis and what our teachers called the *scientific method*. Further along in school, we learned that there were various areas of science: physics, biology, chemistry, and one that seemed quite different—social science. Under this heading, we found fields like history, sociology, and psychology. In college we learned that there were still more subjects and almost all claimed to be in some way scientific: finance, marketing, accounting, woman's studies, diversity studies.

In certain respects, science is all these things. In every respect, it's much more. The world around us behaves in predictable ways. Atoms, molecules, organisms, and human beings display characteristic patterns of movement and association. Science is the tool that over time we've developed to identify and describe these patterns. Physics maps the behavior of matter on levels that range from the quark and the subatomic to the galactic and the cosmic. Chemistry maps the behavior of matter on the level of the atom and molecule. Biology maps the behavior of matter on the organic level. The social sciences map human behavior. Is not the purpose of history to prevent the leaders of today from repeating the mistakes of the past?

To accomplish its role as man and nature's cartographer, science uses a clearly defined methodology. We formulate a hypothesis, or abstraction, that describes how we think some aspect of the world should behave and then test our prediction against what we observe. This link between abstraction and observation is what makes science science. If a model can be tested against the world, it's science. If not, it's not science. This doesn't mean that a model that can't be empirically tested is without value, only that as we define science it's not science.

And, just as there are good arguments and bad arguments there is good science and bad science. Good science establishes a clear relationship between what we think will happen and what we observe to happen. Bad science claims to establish such a relationship but when looked at critically fails to stand up to our scrutiny.

The technique we use to establish a relationship between the model and reality is the study, or experiment. A well constructed study, or one that if the experiment is repeated will give the same results, does two predominant things: It isolates the variables under investigation, and it incorporates reliable methods of measurement.

This is relatively easy to do in, say, a chemistry experiment where we can put the reactants in a beaker or some other piece of lab equipment and weigh the products. So functional is science on the level of physical research that it's hard to imagine a better tool of inquiry. In more complex biological and human systems, designing a good experiment can be a challenge.

Take the testing of a new drug. When introduced into the body, a drug will have many effects. We test a drug to measure these effects and to determine if the ones we want outweigh the ones we don't want. Does say a drug intended to cut down on heartburn in fact cut down on heartburn, and is a decrease in heartburn worth an increase in diarrhea and headaches. The best way to test a drug is with a *double-blind* study. In this type of study, one group of subjects is given the drug and another a placebo, but neither the subject nor the researcher knows which he or she is taking or administering. This isolates the variable of the drug and allows its effects to be measured against a standard.

The science of disease is more complex. Consider the variables associated with heart disease: age, sex, diet, stress, smoking, genetics, exercise, blood pressure, and blood cholesterol to name a few. How in an experiment can we isolate one variable from another and determine which has the greatest influence? Diet and exercise, for example, may offset the effect of a genetic predisposition for heart disease, and stress may offset the effect of diet and exercise. We've all heard of the sausage-loving smoker who lived to be a hundred. We've all heard of the tofu-eating marathon runner who dropped dead of a heart attack at forty. So difficult is it to isolate the variables of heart disease and to measure the results of their manipulation that researchers have thus far been able to conclude with certainty little more than a given variable represents a *risk factor* associated with a higher incidence of heart disease.

Consider also the distinction between anecdotal evidence and empirical evidence. How many times has someone told us that they take a vitamin or supplement and haven't caught a cold in years, therefore the vitamin or supplement prevented their colds. In their minds, they established a cause and effect relationship between the vitamin or supplement and their good health. As our heart disease example illustrated, cause and effect relationships are quite difficult to establish. The person may not have taken a vitamin or supplement, or taken another vitamin or supplement, and still not gotten a cold. Without an element of control, or some way to isolate the variable under consideration and measure an outcome, there's no way to know. From a scientific standpoint, the above statement is no different than the argument that I regularly eat double bacon cheeseburgers and haven't caught a cold in years, therefore double bacon cheeseburgers prevent colds. The latter anecdote is just as valid, and you'll find plenty of people who can make such a claim. It just doesn't sound as healthy. Anecdotal evidence rests on what people believe and may have experienced. Such evidence is valuable in that it can suggest connections and encourage empirical research, but it's not science and should be approached with this understanding.

As hard as it is to map organic behavior, it's harder to map human behavior. A sociological study might determine that lower income families have a higher incidence of drug and alcohol abuse. A reasonable possibility, but did the study take into consideration the genetic factors associated with drug and alcohol addiction? What about diet, education, upbringing, and a host of other variables? As important, what criteria did the study use to measure income and to quantify drug and alcohol abuse? Did lower income lead to a higher incidence of drug and alcohol abuse, or did drug and alcohol abuse lead to a lower income—or to a higher unemployment rate which led to a lower income, or to a higher divorce rate which led to a lower income? So difficult is it to map human behavior that virtually every study in the social sciences qualifies its conclusions by stating that the results warrant further research. Not a bad deal if you make your living off research grant money.

Science is the process whereby we map the behavior of the world around us. Good science establishes a clear relationship between what we postulate and what we observe. Bad science only claims to establish such a relationship. To maintain scientific integrity, an experiment must isolate the variables under investigation and accurately quantify these variables. As systems grow more complex, this becomes increasingly difficult to do. On all levels of complexity though, science can be a useful and powerful tool. But to make or counter an argument supported by science, we must know what science is and keep in mind its strengths and potential for misuse. Invoking the name of science does not science make.

3

Statistics

The soundness of the arguments we make and counter rest on the truth of their premises, which usually rests on science. Science is the tool that over the more than two thousand years of its existence, and the more than three hundred years of its modern practice, we developed to compare how we think the world should behave with how we observe it to behave. To conduct science, then, we need a way to make this comparison. We need an implement of association. Just as we have invented and continue to perfect science we have invented and continue to perfect such an implement—mathematics, which if we're not a scientist, engineer, or mathematician we most often encounter in the form of data, polls, and surveys—statistics.

To understand and learn how to interpret the facts-and-figures shot at us from every imaginable source in support of every imaginable point, we need to know something about math.

First of all, mathematics isn't arithmetic. It isn't the mechanics of addition, subtraction, division, multiplication, roots, and powers we learned in grade school. These skills are the techniques we use to perform math—the syntax and grammar of math. This brings us to math itself. Beyond the mystique of complexity and abstraction, mathematics is a language.

Specifically, mathematics is a system of expression designed to show a relationship between measurable quantities. In English, we would say that boiling water is hotter than tap water. In the language of math, we would say that—if tap water measures 10 degrees Celsius and boiling water measures 100 degrees Celsius—boiling water is 90 degrees hotter than tap water. In English, we would say that if we pass a car on the freeway we're traveling faster. In the language of math we would say that—if our car has a speed of 60 miles per hour and the car we're passing has a speed of 50 miles per hour—we're traveling 10 miles per hour faster. Mathematics relates the abstract world of the mind and model to the quantifiable world of natural and human behavior. Mathematics is the tool of comparison, the language of science.

But like any language, mathematics needn't convey the truth. If when we measure the temperature of tap water our thermometer is off, the calculated temperature difference between tap and boiling water would be incorrect. Results can never be better than the data from which they're derived. It's understood that the universe has four dimensions: back-and-forth, front-and-back, up-and-down, and time. Mathematically, we express these as duration and the X, Y, Z axes of the Cartesian coordinate system. Four dimensions are useful. We couldn't find our way to work or school and get there on time without them. But do they exist or are they mathematically imposed on the world? What about the ten or more dimensions dealt with by the mathematical construct of physics popularly called String Theory? Mathematics is a language. We can use it to describe the world or to impose our conceptions on the world.

This is particularly evident in the area of mathematics called statistics. Studies and experiments often produce large volumes of data. Statistical techniques allow us to discern patterns in this data. A statistic may tell us that something exists and to what degree: Twenty percent of the general population believes that we've developed an HIV vaccine. A statistic may tell us the correlation between two or more variables: As blood pressure increases the number of heart attacks increases. A statistic may tell us the likelihood that something will happen: As blood cholesterol increases the probability of a heart attack increases.

Statistical techniques do a remarkable job of squeezing useful information out of otherwise indecipherable quantities of data. But to be meaningful, we must know how to interpret the information we derive. Take opinion polls and surveys.

In theory, the best poll would survey everyone of interest, an entire *population*. But it's almost never practical to do so. Instead, pollsters survey a sample of the population. If the opinions of those in the sample are proportionate to those in the population as a whole, the poll will reflect everyone's views. A study where the results are derived from a *representative sample*—or one where measures, such as randomizing data, are taken to assure proportionality—is said to be done *scientifically*. Marketing surveys are usually scientific. Corporate decision makers sincerely want to know who is and who isn't buying their products and for what reasons.

Political polls are another matter. A political poll may be scientific—and most claim to be—but may also be designed to yield results slanted, or *skewered*, in support of an agenda. Take a survey conducted on a liberal or conservative web site. The survey will draw its sample population from visitors to the site, which will largely share the views presented on the site and as such give results that support those views. How valid are studies done by a globalization or an environmental group, by an agribusiness or an animal rights group?

Pollsters also skew the results of a political survey by wording and selecting questions to yield answers that support a desired objective or point of view.

Would you vote for President Bush or for a Democratic contender?

This question, asked on a number of polls before the Democratic primaries and the selection of a Democratic candidate in the 2004 presidential race against President Bush, frequently indicated that Bush was running neck-n-neck with his unspecified contender.

Would you vote for President Bush or for Howard Dean?

Asked during the same period but referencing a possible Democratic contender, President Bush often came out ahead by a wide margin, as high as two-to-one.

Given that President Bush would never run against an unnamed opponent, which question is valid? Actually, both were asked so far before the election that neither was intended to gauge the outcome of the race and had any validity at all for that purpose. Howard Dean never even became Bush's contender, that role falling to John Kerry. The questions were worded and selected for a more subtle reason. They sent a message to the electorate that both parties could field a potentially electable candidate. Whatever our opinions, we like to know we're not alone. Were political polls in the Barack Obama versus Hillary Clinton or in the Barack Obama versus John McCain elections used for a similar purpose? How about recent polls in state and local elections?

Mathematics is the language of empirical comparison, the tool we use to conduct the business of science. Is our data sound, our measurements accurate? Does our math bring to light a pattern in the way the world behaves, or are we using our math to impose our conceptions on what we observe? Does a poll or survey claim to be scientific? Does it come from an objective source or from one with an agenda? We read and hear the results of countless studies and surveys, too many to evaluate their methodology even if we had the access and mathematical skill to do so. Should we dismiss all results? One may be inclined to think so, but good surveys and studies do provide useful information. It's our obligation, however, to weigh statistics against common sense.

4

Diet

WHAT WE EAT, WHAT we want to eat, and what we're told to eat is an issue we all face. Meat. Eggs. Fat. Sugar. Starch. Organic and inorganic. Herbal this and herbal that. The volume of dietary information thrust upon us is truly remarkable—a deluge of opinions, research, and statistics. But what about a *common sense diet*?

If we could pick anything to eat, we'd choose foods that tasted good and that left us satisfied, energetic, and steady of mind. We'd choose foods that didn't give us gas and that didn't make us fat or skinny or cause a disease. Does such a diet exist? Yes! And it's common sense. The healthiest diet is the one our bodies were built to consume and as such the one we find most appetizing, the diet we were designed to digest and metabolize.

What, then, does this diet consist of? To determine this, we need to set aside our dietary preconceptions, our nutritional rights and wrongs. Objectively, human beings are *omnivorous*. Unlike a cow, or herbivore, that's designed to eat grass, or a lion, or carnivore, that's designed to eat meat, we're generalists. Our bodies are designed to process a wide range of foods.

We know this from our dental structure. Different teeth have different uses. Front teeth, or incisors, cut and gnaw. Canines bit and puncture. Molars grind and pulverize. Cats, wolves, and other predators have large canines. Deer, horses, and other herbivores have large incisors. We don't have any prominent teeth. Our dental structure doesn't favor the consumption of one food over another but is designed to handle a wide range of foods.

We also know we're omnivorous from the archeological evidence of our evolution. Five million years ago, our early ancestors gathered fruits and tubers and hunted and scavenged game. By about two million years ago, they had made the first stone tools and had used them to scrape hides, extract marrow, and butcher carcasses. By about one million years ago, they had harnessed fire and had begun to cook their food. We evolved as omnivores from omnivores.

Archaeology tells us one more thing about our diet. Unlike our closest omnivorous relative, the chimpanzee—who only on occasion hunts, scavenges, and eats meat—our ancestors routinely hunted, scavenged, and ate meat.

We know this for several reasons: First, evidence of animal consumption is pervasive throughout the human fossil record. Where we find evidence of early humans, we find evidence of hunting, fishing, scavenging, and meat processing.

Second, the nutritional yield of natural food plants, such as the wild varieties of nightshade from which we derived the potato, is a fraction of that of their domesticated counterparts—which is why eleven thousand years ago, a heartbeat in evolutionary time, we began to domesticate them. Like the mountain gorilla, who lives off wild celery, and like every other herbivore, if prior to domestication we had been largely vegetarian we would have been forced to spend almost every waking moment chewing and swallowing. As an intelligent creature, would you rather spend your day digging and gnawing on roots or on occasion hunt and, as does the lion and every other predator, spend your day lounging in the shade?

Third, in all but the lushest environments not enough natural food plants are available to meet our calorie needs. Two centuries ago, the hunting and gathering tribes of the American plains consumed a wide variety of tubers and berries, but their existence depended on and culture revolved around the buffalo. Tribes further north, and in particular those on the fringes of the arctic ice flows, subsisted almost entirely on animal fat and protein. Without a shift from a vegetable to a meat based diet, humankind could never have migrated across the continents.

As a result of our omnivorous nature and five or more million years of evolution, our bodies are designed to work best on a varied diet of raw and cooked foods that includes a reasonable allotment of protein. This is the common sense diet.

Today, we have almost every food available and can choose to eat a commonsense diet. Why don't we? Until the mid 1960s many of us did. Modern agriculture enabled us to toss the bread and potato diet of our peasant forbearers and eat the diet we were drawn to and over the decades found to work best—the varied, protein-centered diet of our more ancient hunting and gathering forbearers. This was encapsulated in the dietary guidelines of what in the United States was called the four food groups, sometimes illustrated by the *food circle*. At the hub of the circle was protein: eggs, meat, fish, dairy, chicken, and their associated fats. Around the hub were sugars and carbohydrates: grains, fruits, and vegetables. Children grew taller. Obesity was rare. Anorexia and other eating disorders were practically unheard of, and life expectancy soared.

Then came greed, politics, and bad science. Researchers saw a correlation between obesity, cholesterol, and heart disease and stretched their logic to conclude that foods high in fat and cholesterol—eggs, meat, dairy, poultry, and shellfish—caused heart disease. Politicians replaced the food circle with the *food pyramid*—a diet high in carbohydrates and, to reduce fat and cholesterol, low in protein. The pharmaceutical industry marketed cholesterol-lowering drugs. The supplement industry marketed herbs and a plethora of nutritional concoctions. Obesity, diabetes, and other nutrition related concerns shot up. Girls reached puberty at a younger age. Doctors diagnosed eating disorders and a "disease" called *Attention Deficit Disorder*.

Though residents of Chicago, which enacted a recently overturned ban on eating goose livers, and residents of New York City, which outlawed the eating of trans-fats and today is attempting to regulate salt consumption, may disagree, the "food-police" don't exist—and won't unless we let them. Despite the government's now more than one dozen food pyramids and hundreds of thousands of pages of nutritional research and guidelines, what we eat is up to us. Our bodies require raw and cooked foods. Whether raw or cooked, we need them in the form we've evolved to digest and metabolize: water instead of soda, fruit instead of pastry, whole grain bread instead of white bread. This doesn't mean we've eaten our last donut or sipped our last soda. These things won't kill us, and let's be honest they do taste good. It means that our body will function better on a diet that includes more unprocessed foods and less refined sugars and carbohydrates, a conclusion that not even the soft-drink and snack-food manufacturers will take exception with.

Our body also requires proteins and their associated fats. Growing evidence, for example, suggests that Attention Deficit Disorder in children may be, at least in part, a symptom of blood sugar fluctuations caused by a low protein, high carbohydrate diet. Many factors have been hypothesized to contribute to Attention Deficit Disorder. These range from diet and exercise to genetics and social environment. There is also no conscientious as to whether Attention Deficit Disorder has a biological basis and as such should be regarded as a disease. Differences of opinion are particularly strong with regard to mild cases and the recommendation by schools and doctors that children be prescribed medications to calm them down. With regard to diet, protein slows digestion and when eaten with carbohydrates stabilizes insulin and blood sugar levels and eliminates a craving for more carbohydrates. Our meals, breakfast in particular, should consist of a serving of protein supplemented with grains, fruits, and vegetables. Attention Deficit Disorder may be a controversial issue, but it goes without saying that children have better focus and do better in school on a full stomach and if they start their day with a good breakfast.

Common sense also tells us that we must adjust our diet to our body's individual requirements. Some people can digest dairy, others can't. If milk gives you diarrhea, think twice the next time you pour yourself a glass of milk. If not, enjoy. Also important, our body has a genetic predisposition to a certain weight range. We're naturally skinny, naturally heavy. We can go above or below our range, but not without a strict diet and possible health consequences. Reducing calorie consumption, for example, can decrease body fat. This may be useful to drop into a genetic range, but may be unhealthy if used to force oneself below—despite what the doctor's obesity chart and its politically motivated and ever dropping ideal weights may say. Our body knows what it needs. We have to listen to it. As important, we eat better and feel healthier when we enjoy our foods. Our meals should taste good and look good. Cooking and serving provide a remarkable creative outlet. Try new recipes. Present a well-set table. We should look forward to our meals and to the company and conversation they provide.

We must also trust our common sense when it comes to nutritional information. Research is useful and is proving the value of a commonsense diet. But much of what we read and hear is hype or outright wrong. Careers and millions of dollars rest on the outcome of research. Every supplement manufacturer claims their product is clinically proven, but where are the peer reviewed double-blind studies? For some individuals, reasonable doses of vitamins may be of value, but no pill can replace food which, in addition to the best known vitamins, contains thousands of micronutrients.

Then there's soy. The health food industry touts it as a gift from Mother Nature. In reality, it was a gift from global agribusiness. Until the 1960s, the soybean was seen as a low-value industrial crop used to fix nitrogen in soil and as a source of oil to make automotive and other lubricants. Unless fermented—the form in which it was traditionally consumed in Asia—soy was hard to digest and considered inedible. In need of an outlet for what was a waste product of the crop rotation cycle, marketers took note of an emerging "green" segment—tofu.

Here we sit on our carbohydrate-loaded asses—manic, overweight, the victims of bad science, good marketing, and do-good politics, our blood surging with supplements and mood and cholesterol altering drugs. But it's our choice. The common sense diet is simple and available. It's what tastes good and what our bodies were designed to digest and metabolize. To be healthy and feel our best, we need a varied diet that is reasonably high in protein.

5

Exercise

When it comes to good health, a commonsense diet is important. So is exercise. And, as with diet, we face a deluge of data, opinions, and research telling us how we should exercise and what videos and apparatuses we must buy. From the standpoint of practicality, how much exercise do we need and of what type? What is the common sense approach to fitness?

To determine this, we begin with anatomy. Our body consists of two types of muscles. The first are the *voluntary muscles*, or those that move our arms, legs, and other major body groups and that we consciously control. The second are the *involuntary muscles*, or those that power our heart, lungs, and other internal systems and that for the most part we don't consciously control. Fitness experts call the voluntary muscles the *large muscles*, or simply the *muscles*, and the category of involuntary muscles most associated with fitness the *cardiovascular system*.

No matter what type of muscle, performance increases with use, or when we contract a muscle against *resistance*. If we do a set of squats, or knee bends, the resistance to our movement created by our body's weight breaks down tissue in our thigh muscles. When we rest, the muscle cells regenerate that tissue. They also do something remarkable. In response to our activity and the tissue breakdown it caused, our cells sense we'll ask them to work harder in the future. They not only regenerate lost tissue but build more. Muscles grow in response to resistance.

Any exercise breaks down muscle tissue and incites growth. When it comes to our large muscles, however, the most effective exercise is *resistance training*. Resistance training, most often done with weights, allows us to isolate a muscle group and by controlling the weight and the repetitions progressively increase the resistance we apply to the group, systematically increasing muscle mass and strength. So effective is weight and other forms of resistance training that a set or two of squats can produce as much muscle breakdown and stimulate as much muscle growth as miles of walking or running. Resistance training is the muscle builder.

But as desirable as it is to build muscle mass and strength, there's more to fitness. We all know the guy in the gym with zero body fat and who can lift more than anybody but who, on the basketball court, can't pivot or dribble. To be fit, we need muscle strength. We also need to develop our ability to control that strength. We need to increase our *nervous system response*. The skier and soccer player lift weights to build muscle mass but also perform sprints, rope jumping, and a variety of drills to increase balance and the ability to at a snap move.

And to bring fuel and oxygen to our nerves and muscles, we must have a healthy cardiovascular system. To strengthen our cardiovascular system, we perform exercises that increase our heart and breathing rate and that allow us to hold that rate for a period of time. We then progressively lengthen the time. This requires exercises where we perform gentle movements over and over. Exercises of this type, commonly called *aerobic exercises*, include biking, walking, running, swimming, and fast-paced dancing.

Based on our anatomy, a good exercise program will have each fitness component: We build muscle size and mass with weight, or resistance, training. We build nervous system response with drills and exercises that improve our balance and our ability to move. We build cardiovascular performance with aerobic exercises.

But if a workout that includes the three fitness components is all there is to good fitness, why are we in such bad shape? Why is it hard to stick to an exercise program?

Most of us think of exercise as something we do in addition to everything else, one more thing to fit into an already hectic day. As such, when our schedule tightens or our energy runs out, exercise is the first thing to go. To stick with an exercise program we must incorporate exercise into our daily activities in a way that adds to rather than imposes a burden on our life.

How we do this depends on what we want exercise to do for us. If we're competing in a sport, exercise will occupy our time in proportion to the level of competition. A weekend volleyball player will train a little. An Olympic contender will train a lot. For most of us, exercise shouldn't be a means to an end. It should be fun, something we look forward to. Do you like to take a morning walk? Do you like to run, swim, or bike?

Do you like to spend time in the gym and hit the weights? Some people enjoy working-out at a health club. Others can't afford a club or prefer to exercise alone. For the cost of a couple months dues, a jump rope, a bench, and a set of free weights will give a terrific workout. We must also look for ways to incorporate exercise into our activities. Take the stairs. Walk the dog. Walk the kids to school. Park the car in the back of the lot. These steps take little time and add miles of walking.

As important as exercise is, however, we can exercise too much. Conditioning takes place through the breakdown and regeneration of muscle tissue. The time it takes for our muscles to regenerate varies. A young person may be able to lift weights five days a week without overdoing it, or *burning out*. An older person may be able to lift weights three or fewer days a week. A well-designed exercise program gives muscles the recovery time they need to reach their maximum size and strength between workouts. If progress levels out, exercise less often or less strenuously. Alternate upper and lower body workouts. Shift the balance between the three fitness components. Most muscle regeneration happens while we sleep. Enjoy a good night's rest. To get the most out of exercise, adjust workouts and expectations.

At the start of every exercise program, we set goals and can't wait to get to the gym. Before long we haven't made the gains we'd hoped for and get bored, frustrated, and discouraged. To avoid this scenario, we need an exercise program that includes the three fitness components: muscle building, nervous system response, and cardiovascular development. And we must incorporate exercise into our daily activities in a way that adds to rather than detracts from the quality of our life. What exercises do we like, and can we blend exercise into our other activities? Do we feel comfortable exercising at home or at the gym? Do we let our body tell us how strenuously and how often to work out, and adjust our exercise routines accordingly?

6

Self

We've looked at diet and exercise, two important and well researched aspects of our *physical self*, but what about our *essential self*? What about that part of us that's reading these words, that dimension of who we are that harbors our common sense? We human beings are the most unique of creatures. What sets us apart from other life? What drives us to do all that we do? This and the chapter that follows, *Collective*, present the most philosophical and thought demanding ideas in the book. They may also be the most important. The chapters establish the insight into our nature that, as individuals and as members of the human community, we need to get to the commonsense heart of topics as far ranging as family, society, politics, socialism, capitalism, and democracy.

For a hundred or more millennia, we human beings have asked what makes us human. And, in our act of doing so, one answer has throughout time stood out. Interestingly, it has nothing to do with morphology. We're not human because we stand upright and walk on our hind legs. We're not human because we have an opposable thumb, stereoscopic vision, and a bigger brain than any other creature on earth. We are who we are because we ask who we are. We are who we are because we ponder the reason for our existence.

Other animals think and learn. Other animals experience love and happiness, fear and loneliness. You and I embody more. We feel a distance in our being, a separation apparent when we converse with ourselves, a sense we are ourselves and at the same time are removed from ourselves. Not only do we think and learn; we know we think and learn. Not only do we experience love and happiness, fear and loneliness; we know we feel these things. You and I aren't merely conscious beings; we're beings aware of our consciousness.

As such, we have the capacity to dwell on our existence. We ponder the origin of the universe. We ask how and for what reason we and all we see around us came into being. The definitive characteristic of the human being is made possible by our anatomy but isn't anatomical. It's our uniquely human ability to look at ourselves and our world as if we were distant from ourselves and our world—the capacity to *reflect* on the meaning of it all.

And, as reflective beings, we're driven to grow and learn. Even a cursory look at history and the events that every day unfold in the world make it clear that our motivation is to move forward, to expand the reaches of thought, to push outward the limits of what it means to be human. We're driven to evolve, to reinvent ourselves, to create within ourselves a higher state of being, to advance to a state of greater insight and awareness. Moreover, as did the twentieth century psychologist Abraham Maslow, who saw human motivation as a progression from basic needs to higher needs of "self actualization," we can categorize the human desire to advance to greater levels of insight and expression. We can plot a hierarchy of needs.

As Maslow understood, practical reality compels us to focus our energy on our material well-being. The first level in our hierarchy of human motivation is *economic*. When hungry, we seek food. When cold, we seek clothing. When caught in the elements, we seek shelter. As well, our material possessions help define our place in society. When we lack the trappings that express our social standing, we're driven to procure the cars, homes, clothes, and other artifacts that illustrate our achievements in life and our relationship to those around us.

But human motivation isn't entirely economic. When we're fortunate enough to reach the point in our lives where we experience the fulfillment of our material needs—and when we're fortunate enough to grasp the nature of these needs and by doing so gain the perspective to look beyond—our motivation crosses the threshold to a higher level of expression. We're driven by a more profound calling, by the need for *freedom*. To some freedom is a concept, a notion we associate with law and politics. To others freedom is a right or a privilege. We enjoy or strive to enjoy freedom of speech, freedom of government, freedom to live and work where we want. But freedom has a deeper significance. It's the environment where we can grow and learn, the situation where we can reinvent ourselves. It's the state of the universe where as reflective beings we can realize our nature. However we may define it, and however we may fight to bring it into our lives, freedom is that which allows human beings to be human.

As profound as our quest for freedom may be—as forceful as it may have been in shaping our lives and history—human motivation reaches a still higher level of expression. Endowed with freedom, we experience the need for *perfection*. We're motivated by the desire to reshape ourselves and our world into ever more aesthetic, functional, and satisfying forms. We work to establish more fulfilling relationships with our family and others in our life. We work to perform whatever we do to the best of our abilities. A nurse strives to interact with her patients in the most satisfying manner. A teacher strives to nurture his students in the most stimulating way. A homebuilder strives to build the perfect house and to make each new house better than the last. We take satisfaction in molding the world within our reach to our ideal of perfection.

As human beings, we experience a hierarchy of needs. Our economic needs fulfilled, we strive for freedom. Empowered by freedom, we strive for perfection. Our hierarchy of needs, however, isn't rigid or absolute. At one moment, we may devote our energy to economic matters, at another to the quest for freedom, and at still another to a job well done.

Moreover, as the centuries and millennia have passed, our human drive to push ahead has elevated our hierarchy of needs to new levels of expression. Modern agriculture and industry have reduced the effort we expend on subsistence. Few of us grow or hunt our own food and build our own houses. We enjoy greater freedom under democracy, with its rights and constitutions, than under fascism or communism, with its dictators and central control, or under seignorialism, with its lords and peasants. The value we place on ourselves as individuals has become more pronounced. In the developed world, few of us can imagine living under the will of a godlike king. Still fewer can imagine owning a slave or of being enslaved. We've become more aware, more independent, more autonomous, more driven to act in ways that are strictly our own. We as individuals have risen to dominate the human experience. Human existence has evolved to accept the individual as unique and of intrinsic value.

Our essential self stands apart from our physical self. Reflective in nature, we're beings who are aware that we're aware, who understand that we understand. As such, we're driven to attain ever higher levels of existence. Given the opportunity to channel our energy beyond economic desire, we strive for freedom. Given freedom, we feel hopeful, vigorous, imaginative, forward looking, driven to better ourselves and our world. The human community evolves through the evolution of the individual. Humanity cannot attain its highest level of expression until every human being attains his or her highest level of expression. Human progress rests on our drive, on our vision, on our quest for the future. At the heart of it all is the "self."

7

Collective

Human progress may rest on the "self," on our individual drive and vision for the future, but that doesn't mean we're meant to spend our lives alone. As human beings, we experience the need to be with others. This brings us to the idea of the *collective*. The word "collective" is tossed around by pundits, often in the phrase "for the collective good" or with regard to discussions on socialism and communism. Based on the loose use of the word, few of us are clear as to what it means, and fewer still grasp the relationship between the individual and the collective.

So, we start our look at the idea of the collective with a definition. In physics, there is a state of matter called plasma. Plasma is a hot, energetic gas composed of ions, nuclei, and electrons in no particular relationship to one another—a random, uniform blend. When the substance cools and leaves the plasma state, however, its components come together to form arrangements. Ions join to form ionic compounds, and electrons are drawn to nuclei to form stable atoms. A substance such as plasma—where the components are arranged uniformly, or with little structure or relationships between members—is said to be "more collective." A substance where the components are nestled—or organized into structural levels that make up successively higher structural levels, atoms that align to make molecules, which align to make more complex structures—is said to be "less collective."[1]

This definition brings us to what is, with regard to collectivity and social organization, the single most important observation we can make, one that allows us to explore the nature of social organization from a par-

---

1. The term "collective" was used more loosely in the past. The French philosopher and definitive writer on the topic Pierre Teilhard de Chardin, for example, used it to refer to the tendency of components to aggregate into more complex arrangements. Our definition reflects the use of the term as it's commonly understood today.

ticularly revealing angle. Clearly seen in life—and, as our plasma example illustrates, with roots traceable to the inorganic—social organization has changed over time. It has evolved in an unmistakable direction. In the nonhuman realm, as more advanced species emerged and, in the human realm, as more advanced societies developed, social structure has become more layered and complex, nestled and interwoven—less collective.

A random distribution of cells in pond water is more collective than the arrangement of cells in a pond slime, which is more collective than the arrangement of cells in a cell colony. In ways we've only begun to understand, bacteria and other primitive life forms communicate through chemical secretions to coordinate reproduction and their relationship to one another. Further up the evolutionary ladder, the marine polyps whose calcareous secretions form coral display a still less collective social organization. A few rungs higher, we find the classic example of a collective society, the ant colony. In a typical colony, infertile worker ants gather food and build and defend the nest, and male ants impregnate the queen whose sole duty is to produce offspring. The colony may be large and demonstrate a division of labor, but the bonds between the members are weak. Similar to what we find in the bee hive, hornet nest, and with other insects, social structure is uniform and mechanistic.

As we continue up the evolutionary ladder, the uniformity and mechanistic quality of social organization decreases. This is particularly apparent on the level of the mammal. In the bison herd, females with young form tight bands and males form their own groups. In the lion pride, adult males, females, and young maintain closely-knit subpride relationships within the highly organized pride structure. The wolf pack displays an even more complex organization comprised of parents and offspring. When we reach the level of the ape and monkey, species such as the gibbon display a rudimentary family and community structure.

The ascent of social structure out of the collective is even more apparent in the human realm. Early hunting and gathering cultures maintained a communal way of life where, with the general exception of group leaders, members had roughly equal standing in society. Over time, groups organized to form tribes, and tribes organized to form allegiances of tribes. Within the tribe and community structure, the family became more important and evolved to become the basis for less collective forms of the community. Communities bonded to form cities. Cities bonded to form states, states bonded to form nations, and nations bonded to form empires.

Today, human social structure is highly nestled and interwoven. We have the immediate and extended family. We have our circle of friends, the community, the city, the state, and the nation. There are international bodies like the World Bank and United Nations, allegiances between nations, and corporate, professional, and other entities that exist as independent social structures within the global political and governmental framework. With respect to the past, contemporary civilization is highly layered and complex—less collective.

Across the spectrum of life, the evolutionary trend from more to less collective social structure is definitive. What, then, is the mechanism behind this trend? Why has social structure evolved to more nestled and interwoven forms? The answer stands out in the human realm, and it centers on the "self."

A social bond is made when we perceive and are perceived as the object of the need to belong and by doing so form a relationship. A wife and husband perceive one another as the object of the need to belong and create a marriage. Parents and offspring perceive one another as the object of the need to belong and create a family. A lord and his peasants perceive one another as the object of the need to belong and create a duchy. A king and his subjects perceive one another as the object of the need to belong and create a kingdom. Lords must have their peasants and peasants their lords. Kings must have their subjects and subjects their kings.

Furthermore, the stronger the social bond we're able to form, the stronger the relationship we're able to create and the less collective the social structure that results. In a communal hunting and gathering society, every member, with in most instances the exception of a ruling elite, has roughly the same social standing and is bonded to every other member in roughly the same way. Social structure is uniform and egalitarian—more collective. In a modern society, a husband and wife bond to form a couple, which bonds with its offspring to form a family, which bonds with other families and with individuals to form a circle of friends, which becomes the basis for a community of associations. Social structure is layered and complex—less collective.

As we rose to a greater sense of "self"—as our hierarchy of needs reached new levels of expression and the individual became increasingly dominant in the human experience—our social relationships evolved in response. As you and I attained greater autonomy and substance of character as individuals, the social bonds we formed became stronger and the social structures we created became less communal and centrally controlled—less collective.

Of late, the term "collective" has become a buzzword on talk-radio and other forums. As presently defined, it has a specific meaning. A more collective society has a more uniform structure and weaker bonds between members. The individual is less important and there is greater dominance by a ruling elite—a greater degree of top-down, or central control. A less collective society has a more nestled and interwoven structure and stronger, more intimate bonds between members. The family and individual are more important and there is less central control. The greater the freedom and autonomy of the individual, the more layered and complex the society. As the "self" became more pronounced, we formed stronger bonds between one another and created less collective societies.

8

Family

Throughout history, our human desire to align with those around us has taken many forms. At the heart of all, however, is the pattern of social bonding that we call the *family*. But, as our previous chapters titled *Self* and *Collective* suggest, and as anthropology makes clear, family life as we think of it today hasn't always been around. The family has changed over time. It has advanced, following an unmistakable line of evolutionary ascent, one that allows us make certain commonsense observations about the contemporary family and where it's headed.

As nomadic hunters and gatherers, we lived communally. We had marriages and created families, but the bonds that defined these relationships weren't as strong as today. Men often had more than one wife, and on occasion women had more than one husband. The group rather than the parents took primary responsibility for child rearing. Couple and family bonds existed but were weak, overshadowed by the social dynamics of communal life.

As time passed, polygamy became less common, and marriages were less often arranged. The couple rather than the group took on the major duties of child rearing. Within the communal social fabric, the family differentiated to become a cohesive unit—the social structure on which we would build less collective communities and go on to create the city, state, empire, and modern nation state.

Today, the center of the family, at least the idealized version we think of as the traditional family, is a man and woman in a lifelong, monogamous relationship. Bonded to the couple are its offspring. This arrangement, called the *nuclear family*, is entwined with the connections to aunts, uncles, nieces, nephews, cousins, and grandparents that make up the *extended family*.

The human rise from the collective—our evolution from communal to family based social structures—is a pervasive trend in the ascent of civilization. As such, it invokes a question. If the trend is toward more intimate, family oriented social arrangements, why does the family of today seem threatened? Divorce is high. Child abuse and neglect are common. Single men and women and same-sex couples are choosing to have or adopt children. In nearly every developed nation, fewer people are willing to commit to marriage, and birthrates have plunged. The nuclear family faces many challenges. In the minds of some, its time has passed.

The evolutionary trend from the communal to the nuclear family is definitive, but it's not without ups-and-downs. To a remarkable degree for cultures of the day, Athens in Ancient Greece embraced the nuclear family. At the same time, neighboring Sparta was the extreme of collectivity. For the greater social good, babies not meeting Spartan standards for health and beauty were killed at birth. Boys deemed worthy of life were separated from their mothers at an early age to train with the men as warriors. Girls remained in a female caste that ran many of Sparta's daily activities. So rigid was this division by sex that marriages took place largely for reproductive purposes, and men and women fulfilled their intimate needs within their respective sexual castes.

Sparta collapsed from within and fell into obscurity; Athens and the ideas of family and freedom that it nurtured lived on to reshape the world. The evolution of today's family to stronger, less collective forms continues, but it has met a bump in the road.

In the minds of many, the greatest threat to the family is economic. Measured in inflation-adjusted dollars, certain segments of the middle class in the United States have by some measures experienced a steady decline in living standards since the mid 1970s. Today, family-wage jobs are scarce and both parents usually work, sometimes at two or three jobs to maintain the standard of living achieved by a single wage-earner in the 1950s and 1960s. Wives and husbands spend little time together, and to a greater extent than in the past children raise themselves. The stress of economic life and our preoccupation with money and the status it represents too often takes precedence over the family relationship.

Closely associated with economic factors is the view within society that the family is less important than it once was, manifest through a social environment and a legal and political structure that devalue the family.

As is widely reported, the United States has the highest per capita imprisonment rate of any nation. The bulk of those imprisoned are young males, in particular African Americans. In certain communities, this has led to a shortage of males available for marriage and, as important, able to take on the responsibilities of marriage. In the larger sense, we as a society place less emphasis on the male and on male responsibility than in the past. We no longer instill in the male the value that it is the role of the man to accept the responsibility to create a stable family unit. As at no time in our history, women find themselves the sole center of family life, raising children in a social and economic environment where maintaining a stable family is difficult.

Conversely, legal and political measures can in some instances make it difficult for the man to exercise his responsibility and contribute to the family, in particular in the circumstance of divorce. Child custody laws and court decisions may not always provide the best situation for children, and the difficulty of balancing personalities, intense emotions, work schedules, transportation issues, and the demands of day-to-day living make visitation arrangements difficult. All too often, fathers with the best intentions play a peripheral role in the child's life. In the mind of the child, dad may be little more than a friendly weekend visitor.

Also significant are the unintended consequences of social programs designed to assist the family. Many welfare and social service programs incentivize work over family. Is it best for a single mother to place her young children into daycare, and work what may be a menial job, or to devote her time to raising her children? How much can a single parent do? In the long run, what is least costly for society? On the flip side of the coin, welfare funds may be contingent on single parent status and, as such, discourage marriage. Certain social programs are of clear benefit, and many welfare to work programs—demonstrated by pilot projects implemented following welfare reform in the 1990s—have moved women to jobs and as importantly prevented them from going back on welfare. To an extent, however, social programs have superseded the role of the family provider, established a cycle of government dependency that in some instances has been passed from generation to generation.

From a broader perspective, the declining value we place on the family rests on a redefinition of the family, in particular in the media and academia. Throughout history, the family has been such a driving force that religion, Christianity specifically, has addressed it in its doctrine.

Dissatisfied with, or alienated from, Christian and other conservative teachings, many dismiss the family as ideological in nature and choose to think of it in contrary ways. No longer must the ideal family consist of a wife, a husband, and their children. It can be a single mother or father, or a gay or lesbian couple. In the minds of many, no form of the family is intrinsically better than any other. So defined, the family is vulnerable to a need we all at times feel and that has become a social norm—our drive for self-gratification. We want that job, that car, that house, that man, that woman, that trip abroad. We choose to have or adopt a child because we want one, not because we make a commitment to create a family. As individuals, we are too often without an anchor, adrift in society, our immediate needs at the forefront of our thoughts and ambitions.

There are those who will argue the point, but the nuclear family is the most functional family arrangement we've evolved. History tells us this. Common sense tells us this. Study after study tells us this. Our understanding of humankind's ascent out of collectivity tells us this. The present decline of the traditional family isn't a trend or a new direction in evolution; it's a setback. As such, how do we revitalize the family?

To strengthen the family, we must begin by accepting what we've just stated. By virtue of its evolution, the family is the fundamental unit of modern civilization. What social structure other than the family can fulfill this role? The traditional family isn't an institution grounded in church and ideology defined by the buzz-phrase of "family values" or any other political notion—liberal or conservative. The roots of family stretch far back in time, further than those of any modern religion or political bent. The strength of the family, in turn, rests on the strength of the individual—not on our ego or on our need to place ourselves at the center of all that happens—but on our vigor, creativity, and substance of character. To build a marriage, a wife and husband must provide for and assist each other in their personal growth. To build a family, parents—traditional or nontraditional—must position themselves as the central influence in their children's lives. A sound family is made up of sound individuals.

We must also address the day-to-day concerns that make family life difficult. Why is it that today's family must have more than one income to meet its financial obligations? Why is it that we have what may be unrealistic expectations concerning the type of home we live in and our other material trappings, the compulsion as individuals and as a nation to live beyond our means? Why is it that we need social service programs to aid the family, and why is it that these programs may not successfully accomplish this goal? At present, a social net is to some degree necessary, in particular with regard to children. Should we not, however, structure this social net so that we don't perpetuate a culture of dependency on government and its services? Though the single parent household may not be the ideal family situation, forty percent of all children in the United States are born to a single mother, and most single mothers work. Mustn't we assure families of safe and affordable daycare and health services?

From the standpoint of theory, we may judge the traditional family to be the most functional family arrangement, but it isn't appropriate to carry this judgment to the level of the individual family. Family arrangements other than the nuclear family aren't by definition "bad." A loving single parent will create a more nurturing family than an unstable or abusive couple. Many gay couples will do a better job raising a child than many heterosexual couples. The nuclear family is without question the ideal family structure and should be nurtured and encouraged, but it isn't the only family structure that can be successful. Nontraditional families form a significant sector of the American population, and we must address the real-world circumstances imposed by this situation. When conditions demand, we must look beyond absolutes and, with an eye on common sense and the best possible outcome, create the norms and conditions that enable us to direct our energy to family life.

The value of the family is anchored in millennia of practice and evolution. Its form has and will continue to advance, and we must be open to this progress—nurturing the family to ever more intimate and satisfying arrangements. But today's family faces a setback in evolution. The family has deteriorated because we as a society and as individuals—for a plethora of reasons and faced with a plethora of circumstances—have let it. It'll regain its vigor because we as a society and as individuals accept the value of the family and the goal to strengthen it. The notion spouted in certain intellectual circles that the traditional family is past its time and that we must accept its demise is shortsighted. By our values, actions, and choices, the nuclear family of today will evolve into the more intimate immediate and extended families on which we'll build the societies of tomorrow.

9

Society

In the last chapter, we saw that the family has evolved from more uniform, or collective forms, to more strongly bonded, or less collective forms. As we previously discussed, society has also evolved from more to less collective forms. The empire of Ancient Sumer rose out of band and tribal allegiances. In Ancient Egypt, citizens took their pharaohs to be gods, all-powerful over their domain and disciples. In ancient Greece and Rome, the family and individual became more important but were subservient to armies, consuls, and senates. Empires gave way to nation-states. Kingdoms gave way to democracies and representative governments. The greater the individual sense of "self," the stronger the social bonds the individual can form and the less collective the society that results—an observation that will allow us to explore with new insight the nature of social change today.

As with the evolution of the family, the ascent of civilization hasn't been without its ups-and-downs. In Ancient Greece, the militaristic city-state of Sparta that we talked about in the previous chapter dominated the region and its neighbors for generations only to—when confronted with a plunging birthrate and rebellions among its slaves and commoners—face defeat at the hand of the Thebans, fall into decline, and become a historic curiosity. Once ruled by Sparta, Athens rose to nurture many of our most cherished ideas of family, freedom, and self-rule.

Formed in 1922 after the Russian revolution, and led by an elite ruling class until its fall in 1991, the Union of Soviet Socialist Republics, also known as the USSR or Soviet Union, embraced a highly controlled, communal way of life. Children were raised as much by the state as by parents, and members of the proletariat, or working class, had roughly the same social standing. So uniform and mechanistic was Soviet society it was likened to that of a beehive.

A few years ago, a retired couple one of the authors met described their 1960s experience of living in a commune. In the commune, each member was to share the work of growing the food and raising the children. As with similar experiments in the 1960s, infighting broke out among the members and the commune fell apart. The retired couple had learned that a communal way of life didn't work, but they didn't understand why. We must cooperate in life, but a society that mandates cooperation restricts our individual creative spirit and we turn against it. The communal experiments of the 1960s were an attempt by people in modern times to adopt a social structure that reflected the values, beliefs, and view of the world held by people millennia ago.

Defined by the turmoil and struggle that is the ascent of civilization, our drive to greater freedom, individuality, and the less collective social orders they support is definitive but hasn't been without trial. In the end, however, the rise from collectivity has prevailed.

But will it today?

We can think of social evolution as influenced by two opposing forces. The human need to advance that we spoke about in our chapter titled *Self* drives us forward in time, to explore new ideas, to invent new ways to do things, to create more nestled and intimate social arrangements. On the other hand, a need we all on occasion feel—the desire to maintain the status-quo—encourages us to look back in time. We're motivated to value the less evolved. We seek solace in the collective social ethics of the past.

We all struggle to accept new ideas and new ways to do things, and prudence in doing so is an admirable quality. Yet, we at times reach a point where our attachment to old ideas and old ways of doing things hinders our development. We arrive at this state when we direct our inventiveness and creative energy to justify ways and ideas for no other reason than they may once have been worthwhile. We call the process whereby we direct our natural ambition and creative power to maintaining the old beyond its usefulness *stagnation*.

Within the individual, stagnation reveals itself as insecurity and as the ego that results when we too tightly associate our sense of self with beliefs kept alive beyond their time. With nothing new or worthwhile to say, a politician will attack his opponent. Within humanity, stagnation reveals itself as dogma, racism, ethnocentrism, and fundamentalism. Is not the turmoil that grips the Islamic world a battle between modern and ancient religious interpretations?

Many philosophers have reduced the story of human conflict to a war between past and future. Many have interpreted the ascent of civilization as a struggle between the novel and the stagnant, the creative and the status-quo, the road to tomorrow and the quagmire of the present.

Is this the struggle we face today?

We all want to create a more peaceful, cooperative world. We can achieve this most noble of objectives in two ways: First, we can mandate cooperation through government or some form of central control. This approach is predicated on the belief that the individual is not capable of directing his or her life for the common good and that an "enlightened elite" must take on the responsibility. Such an elite may be motivated by the need to better humanity. Such an elite may also be motivated by the need for power. Propaganda aside, how many socialist revolutions have turned out to be for "the good of the people"? Second, we can establish the environment in which cooperation will naturally occur—one of freedom and individuality. The greater our sense of "self," the stronger the social bonds we're able to form and the less collective and more cooperative the society that results. When social order is imposed through central control, from the top down, it's transitory. What dictatorship has withstood the test of time? When social order emerges from the bottom up, when it's based on freedom and individuality, it's stable, intrinsic, an outgrowth of social evolution and the foundation for continued social advance.

Today, there are those who seek to create a more peaceful, cooperative world by turning back the clock, by imposing social order from the top down—by creating a more collective society. This objective manifests in many ways: extreme religious or other form of ideological control, extreme economic regulation, and extreme land-use, environmental, and other regulation. In the classic sense, our quest to establish the society of our communal dreams rests on the notion of *egalitarianism*. In modern civilization, social position is caught up with economic position. We live in a world that is driven by competition, motivated by the desire for profit and control over capital. Thus, to create a less economically divided society, there are those who seek to create a socially undifferentiated society. Social position may be related to economic position, but it isn't necessarily the same thing. As individuals, we are of equal worth. In society, no social position is intrinsically better than any other, but every social position is different. We as individuals are members of a family, which forms the core social structure of the community—which, successively, forms the core social structures of the city, state, and nation. Those who call for an egalitarian society mistakenly attempt to solve the problem of economic inequality by establishing social homogeneity. With the goal to eliminate economic class, they seek to revert to a more collective, less family and individual oriented social form.

Dominant in such a form of social organization is government. We lobby for entitlements and socialistic wealth redistribution, for extreme environmental controls and for extreme regulations on business and economic activity. We demand politically correct dress, speech, thought, and behavior. We track each other's activities and demand that the individual conform to collective values and expectations. We encourage schools to take on more of the responsibility of child rearing and encourage parents to take on less. Many proclaim the need for a world government—a global level of central control. Representative of the international community, the United Nations calls for world taxes, courts, and regulations and attempts to overrule the sovereignty of nations.

Do we, today, face a war between novelty and stagnation, creativity and fundamentalism? Do we today take sides in a battle that rages between those who fight to create the less collective social orders of the future and those who fight to bring back the more collective social orders of the past? To be enduring, must not any future global social structure rest on the foundation of a strong individual, family, community, city, state, and nation?

The greater the sense of "self" that we experience as individuals the stronger the social bonds we're able to form and the less collective and more intimate and cooperative the society that results. The labels of Democrat and Republican, traditional and progressive, liberal and conservative aside, in business, academia, and government, with regard to issues as far ranging as science, politics, religion, terrorism, economics, environmentalism, and ethnic conflict, does not human progress and social transformation come down to a struggle between the values of freedom, creativity, and individuality and the values of conformity, government, and central control?

10

Civilization I

THE RELATIONSHIP BETWEEN SELF and collective, family and society is clearly evident in the rise of civilization. The history of civilization is also frequently cited in political and other arguments, including many made and discussed in this book. How many times have we heard a pundit quote Plato or Aristotle or refer to the Crusades in the context of present-day Islamic aggression? Most of us studied the rise of civilization in high school or college, and most of us can benefit from a brief refresher. In this chapter, we trace the ascent of civilization through Ancient Egypt. In future chapters, interspersed with those written on contemporary issues, we pick up where we leave off and highlight our advance from Ancient Greek to modern times.

Civilization, from the word "city," is generally considered to have begun with the human move from a nomadic, hunting and gathering way of life to an urban way of life, which started to take place about twelve thousand years ago. Traditionally, anthropologists felt urbanization began in the *fertile triangle*—the crescent that extends from the Nile Valley to the Tigris and Euphrates rivers in present-day Iraq—then diffused throughout the Near East, spread north into the Indus Valley and China, westward into Europe, and across the Bering Strait into the Americas. Today, most scientists feel that urbanization arose independently at these locations. In either view, the first settlements were little more than areas where people stayed for extended periods during their wanderings. These places may have been caves or clusters of shelters built near a stable source of food and water or near an area with religious significance. As time progressed, life increasingly revolved around these settlements. As humanity rose from the collective, people felt the need to be closer and chose to make the settlements their home.

Urbanization may have originated independently around the globe, but the Middle East occupied a central position in its ascent. One of the oldest urban centers was the *Neolithic*, or New Stone Age, settlement of Jericho. Built around 8000 BC on the present-day West Bank, it had a population of about 2,500. By about 5500 BC, the Fertile Crescent and in particular, *Mesopotamia*, the region of the crescent between the Tigris and Euphrates, was dotted with towns and villages. From this nexus rose the most expansive and well established of the world's early civilizations, *Sumer*. The Sumerians built walled cities and developed writing and basic mathematics. Perpetually at odds with one another and outside peoples, city-states battled using primitive chariots and with armies of soldiers clad in copper armor and wielding copper tipped spears.

As civilization prospered in early Sumer, a far greater civilization rose from a scattering of settlements in the Nile Valley—Ancient Egypt. Our historical account of Ancient Egypt was founded by *Manetho*, a third century BC Ptolemaic priest, who described early Egyptian life in terms of ruling lineages, or dynasties, which are lumped into four major periods: the Early Dynastic, or Archaic period, and the Old, Middle, and New kingdoms.

During the Archaic period, which dates from about 2925 BC to 2575 BC and includes the 1st, 2nd, and 3rd dynasties,[1] the climate was more lush than today. Evidence of settlements ranges from the Nile delta into what are now the low desert regions of Sudan and Southern Egypt. The Archaic period saw the Empire under the rule of a line of as many as thirty kings, who at Saqqara and elsewhere built clay-block mortuary structures, more sophisticated than older burial mounds and predecessors of the pyramids.

The capital of the Old Kingdom, which dates from about 2575 BC to 2130 BC and includes the 4th through the 8th dynasties, was at Memphis. Under the rule of absolute monarchs, taken to be gods, the Egyptians built the first monumental stone structures. These included the Step Pyramid at Saqqara and the pyramid complex at Giza. The Egyptians also excelled at painting, sculpture, navigation, bronze work, and jewelry making. They had a functional knowledge of surgery, antiseptics, and the circulatory system. As astronomical observers with a grasp of geometry and arithmetic, they developed the first 365-day calendar. The Old Kingdom ended

---

1. Some Egyptologists also include a "0" dynasty, which represents the period's first 150 years.

with a decline in the power of the monarchs and their central bureaucracy and a reemergence of city-states.

The Middle Kingdom, which dates from the end of the Old Kingdom to about 1550 BC and is often broken into periods, includes the 9th through the 17th dynasties. In Mesopotamia, the *Akkadians* conquered Sumer and adopted Sumerian culture only to be incorporated into the *Babylonian* empire; and, in Egypt, powerful regional leaders rose in Thebes and Memphis. As a result of this division, art became provincial and no truly monumental works of religious architecture were constructed. The 12th dynasty, however, marked the reunification of Egypt under Amenemhet I and a line of successors, and ushered in another golden age of art and literature. As before, however, the empire's fortunes would be up and down. On the move out of western Asia, the *Hyksos* took control of the Nile delta, opened it to *Phoenicia* and *Palestine*, and initiated a period of turmoil that lasted more than 200 years. During the 17th dynasty, the Theban ruler Kamose and his brother Ahmose I drove out the Hyksos and once again reunited Egypt.

The New Kingdom, which dates from about 1550 BC to 1070 BC, began with the reestablishment of a powerful military and central bureaucracy and with women gaining influence in Egyptian society, earning such titles as royal wife and royal mother. After Ahmose I, a line of expansionist leaders—including Amenhotep I, Thutmose II, and Amenhotep II—militarily extended Egypt's boundaries, conquering Syria, Palestine, and Nubia, a region in North Africa in present-day Sudan that would give rise to the *Nubian* civilization. The diplomacy of Amenhotep III maintained the balance of power among Egypt's neighbors and established an era of peace, art, and architecture. Egypt's polytheist religion, or one with many gods that have different powers and spheres of influence, however, faced a challenge. Amenhotep's son, Akhenaton, married to the legendary first wife Nefertiti, founded a monotheistic religion based on the idea that the sun god Aton was the sole, all-powerful, all-knowing force in the universe. In honor of Aton and to demonstrate the power of his god, Akhenaton abandoned the Egyptian capital of Thebes and built a new capital, Amarna, the horizon of Aton. Nefertiti devoutly followed her husband and his god but in time fell out of favor and was replaced by one of her six daughters, Meritaten.

Akhenaton's successor and son-in-law,[2] Tutankhamen, known for his gold laden tomb discovered in the Valley of the Kings by the British archaeologist Howard Carter in 1922, renounced the universal power of the sun god Aton and, prior to his death at the age of nineteen, returned Egypt's capital to Thebes. Tutankhamen was succeeded by Ramses I, his son Seti I, and Ramses II. Known as one of the greatest of ancient leaders, Ramses led campaigns against Syria, Palestine, the Libyans, and the Hittites. Ramses also led Egypt into a period of monumental construction. He built Ramesseum, or the mortuary temple at Thebes; Abu Simbel, or the hypostyle hall in the Temple of Amon at Al Karnak; and, with enslaved Hebrew labor, the city of Pi-Ramses on the Nile delta.

After the death of Ramses II's successor, Ramses III, the empire entered a period of decline that lasted centuries and centered on the loss of regional influence. In the mind of Manetho, Egypt's ancient period ended in the 30th dynasty when, after a succession of Egyptian and foreign rulers, Egypt was conquered in 322 BC by a civilization and great warrior from across the Mediterranean, the Greek *Alexander the Great*.

---

2. Some Egyptologists believe that Tutankhamen was Akhenaton's son-in-law. Others believe that he was Akhenaton's biological son by a wife other than Nefertiti.

11

Scarcity

IN THIS CHAPTER, WE shift our focus to a topic that impacts our lives in an immediate way—*economics*. The word "economics" brings to mind many thoughts. We reflect on our struggle for food, clothing, and shelter. We ponder budgets, politics, deficits, surpluses, and conflict between nations and classes. Most of us, however, don't know as much about the subject as we may think, presidents and legislators on the top of the list. In particular, we don't understand the basics of economic function. Beneath the trappings of ideology, what is socialism? Beneath the veil of profits and losses, what is capitalism? To get to the commonsense heart of the topic, we begin with the fundamental assumption on which contemporary economic theories are based—*scarcity*.

This returns us to the origin of civilization and to the first permanent settlements. Prior to urbanization, human bands had occupied the same basic ecological niche for millions of years. As nomadic hunters and gatherers, we maintained a relationship with our environment similar to that of any other species in an ecosystem. As does say a baboon troop, we filled a role in the biosphere, and the biosphere provided us with the food and other resources we needed. Groups may have experienced hunger when climates changed and ecosystems realigned, but for the most part the plains and jungles provided for their needs. All that we needed and desired was there for our taking.

When we moved into permanent settlements, however, we broke away from this environmental relationship. Only a limited amount of game could be hunted and fruits and tubers gathered within a reasonable distance from the village. As a result, the early urban community experienced a chronic need for food and other goods. It existed in a state of intrinsic *scarcity of recourses*.

Scarcity of resources, in turn, spurred the development of systems to coordinate the way people gathered food and other goods in the countryside, where they existed in surplus, and transported them into the village, where they existed in deficit. Urbanization founded economics as we typically think of it today—*scarcity-based* economics. As any economic text will tell you, contemporary economic philosophy is based on the notion that resources are by nature limited and human material wants are by nature unlimited and that some mechanism must exist to encourage the production of goods and services and to regulate their allocation.

At its essence, a scarcity-based economic system is a way to regulate human behavior. It's a set of rules and guidelines that, based on the supposition that resources are limited and human material wants are unlimited, assigns tasks and responsibilities and establishes how the goods and services brought into existence by these tasks and responsibilities are divvied out. A scarcity-based economic system is an agreed upon way to conduct subsistence activities based on a mutually accepted view of the world.

How, then, did the population of the early urban community regulate the behavior of its members? To some extent regulation took care of itself. Faced with cold and hunger, the hunter is motivated to pick up his spear, the toolmaker is motivated to chip a spear point, and the builder is motivated to raise a shelter. When someone needed something, they made, took, found, picked, killed, or traded for what they needed. To a degree, regulation took place on its own—through supply and demand, through what we today would call *market forces*.

Market forces alone, however, weren't enough to regulate the economy of the early urban community. Under the dictates of the market, some members of the community—say those who were skilled hunters—would have plenty to eat while others would go hungry. Left to the dictates of supply and demand, some members of the community would have more than they needed and others wouldn't have enough. This disparity of wealth would create tension, and the social structure of the community would lose cohesiveness. To maintain the integrity of the community, and thus to further humankind's social evolution, everyone's needs had to be met, at least to the standards agreed on by the community. There had to be rules to direct individual behavior for the common good. For the community to be a community, some form of *central control* was also necessary.

Central control, of course, had its own consequences. We can imagine communities where everyone gathered by the village fire at night to figure out what needed to be done the next day and to assign the tasks to those best suited or to those who wanted to perform them. This situation certainly existed, but history and archaeology suggest that it wasn't generally the case. In a regulatory system, someone must make the rules. As communities grew larger and the need for resources grew more pressing, it became difficult for the community as a whole to perform this task. An element of the community was given, or more realistically took upon itself, the responsibility to figure out what was needed, who did what, and who received the benefits. Urbanization and the scarcity it created brought into being a *governing class*.

Those who made the rules, of course, had to have a way to make everyone else abide by them. To some extent, this took place through the social structure of the community. We accepted the rules made by our community leaders because we respected them and believed that they had our best interests at heart. But, as we see in the dynamics of any group, rarely does everyone feel their contribution is essential and appreciated. How many groups have we been in where every member felt that the people in charge were running things the right way? In the early urban community, those in power had to have a way to execute their power. In addition to a ruling class, scarcity-based economics required an *enforcing class*. Urbanization founded the rudiments of what we today would call the *police state*.

The emergence of a ruling and enforcing class, of course, brought with it the emergence of the class of individuals of which most were members. The ruling class governed. The enforcing class enforced. But, in the pure sense, neither group directly contributed to the material well-being of the community. They may have implemented and enforced the procedures that allowed economic activity to function more efficiently, but they didn't perform that activity. They didn't gut and skin the antelope. They only ate its meat. With the development of a ruling and enforcing class, there emerged another division in the economic fabric. Most in the community found themselves members of the *working class*.

Scarcity was a fabrication created by ecological isolation; but, from the vantage of those who lived in the early settlements, it was the natural state of existence. Urbanization led to systems of economic activity based on the idea that resources are by nature limited and human material wants are by nature unlimited. In the larger sense, urbanization and the scarcity-based economic systems it spawned transformed our view of the world. For the first time, we saw ourselves as in opposition to our environment. Our lives were a struggle to eke out of the earth what we needed to feed, cloth, and shelter ourselves and our community. Urbanization and scarcity brought into human consciousness the concept of nature—the idea that we are separate from the natural world. Urbanization and the scarcity-based economic systems it spawned also established a play between market forces and central control and led to social structure that was to a greater extent than in the nomadic community defined by economic class. Urbanization and the scarcity-based economic systems it spawned nurtured contemporary ideals of survival and competition, of government, wealth, and property—of man against nature and man against himself.

12

Civilization II

Ancient Egypt is central to our understanding of civilization's ascent, but there were other cultures in the Egyptian timeframe. By the twenty-first century BC, the Mesopotamian city of Babylon had grown in importance and given rise to the influential *Babylonian Empire*. By the twentieth century BC, the Hittites had settled Anatolia, or Asia Minor, and formed the *Hittite Kingdom*. By the sixteenth century BC, Aryan tribes out of India had settled present-day Iran to—led by one of the greatest military leaders of all time, Cyrus the Great—in 546 BC conquer *Lydia*, in present-day Turkey, and in 539 BC conquer Babylon and establish the vast *Persian Empire*.

According to legend, Hebrew culture—which led to Judaism, Christianity, and Islam—began with thirteen tribes united by Abraham in 2000 BC. In the Old Testament books Exodus and Deuteronomy, Moses, in 1200 BC, led the Hebrews from Egypt—where they had been enslaved by the pharaoh, Ramses II, to build Pi-Ramses—to Canaan along the Jordan River. In the mythological account, Moses parted the Red Sea to escape the Pharaoh's army and climbed Mount Sinai to receive the Ten Commandments from God. Under Moses' successor, Joshua, the Hebrews conquered Palestine and during the reign of King David defeated the *Philistines* and made Jerusalem their capital. After the death of David's son Solomon, the tribes split into two kingdoms, Judah and Israel.

Between 3000 BC and 1100 BC, *Minoan* civilization thrived in Crete. Mythologized in the ninth century BC by the Greek writer Homer, Minoa was described as a rich island surrounded by a "wine-dark" ocean. On the island was a great city built over a labyrinth ruled by a creature that was half-bull and half-man called the Minotaur. About 2500 BC, the *Indus Empire*, in its day second is size only to the Egyptian Empire and with cities with running water and functional sewage systems, rose along the Indus River in modern-day India and Pakistan. Thousands of miles away along the Muang He River in what is now China, the *Shang Dynasty*, a sophisticated civilization with writing and metallurgy, prospered from about 1700 BC to 1100 BC.

Early Greek civilization is traceable to 2000 BC when barbarian tribes established the city of *Mycenae*. In 1876, the German archaeologist Heinrich Schliemann discovered the ruins of ancient Mycenae and unearthed royal tombs and lavish palaces. Most of all, we know Mycenae from Homer's *Iliad* and *Odyssey* and their tales of the Trojan War, gods and monsters, the beautiful Helen of Troy, and the murderous, adulterous, and adventurous lives of Electra, Orestes, Achilles, Odysseus, Agamemnon, and Clytemnestra. Mycenaean civilization reached its apogee with the conquest of Crete in 1600 BC and, after wars against Iron Age Invaders from the north between 1200 and 1100 BC—which may have been the inspiration for Homer's Trojan war—fell into decline, plunging Greece into a centuries long dark age with little evidence of the arts, reading, writing, and mathematics.

From its dark age, Ancient Greece rose into its archaic period followed by its classical period, 750 BC to 323 BC. This time span opened on a countryside dotted with city-states, many of which had established colonies as far away as Sicily, France, and the Black Sea. But the defining characteristic of Ancient Greece wasn't its scope or affluence; it was its politics. In many Greek cities, oligarchies of aristocrats overthrew tribal monarchs. Disgruntled aristocrats—backed by slaves, peasants, and a class of wealthy merchants—then overthrew the oligarchies. Called tyrants because they had illegally seized power, leaders such as Gelon of Syracuse, Periander or Corinth, and Polycrates of Samos were in practice wise and popular rulers. Under their leadership, trade, industry, and the arts flourished. Greek philosophy took root with the "speculations" of Thales, Anaximenes, and Anaximander. The greatest cultural shrine became the sanctuary of the oracle at Delphi, and the Greeks flocked to the Olympian Games, first held in 776 BC.

The greatest strength of the tyrants, their benevolence, however, established the conditions for their decline and the emergence of a yet more open governmental form—democracy. By the sixth century BC, Athens and Sparta had become the dominant cities in Greece. Sparta, the collective society we spoke about in earlier chapters, achieved power militarily and maintained rigid control of its slaves, citizens, and conquests. At the same time, the absolute power of the Athenian rulers was on its way out. In 621 BC, the statesman Draco codified Athenian law, which limited the judiciary power of the nobles. In 594 BC, the statesman Solon reformed Draco's rigid, or Draconian, codes and gave citizenship to the lower classes. Under the benevolent rule of the tyrant Pisistratus, Athenian government adopted democratic elements of discussion and participation. His sons and successors, Hippias and Hipparchus, however, reversed these reforms. In 510 BC, this led to an uprising and a democratic victory under the statesman Cleisthenes, who enacted a constitution based on democratic principles.

Empowered by a new social openness, Athens entered one of its most significant periods of artistic and intellectual achievement. But this flowering of creative expression was tempered by war as Ancient Greece confronted its greatest military challenge.

When across the Mediterranean, Cyrus the Great, King of Persia, overthrew Lydia, he brought Greek cities in the coastal islands under Persian control. Backed by Athens and Eretria, many rose against Persia in the Ionian revolt. In 492 BC, Cyrus's successor Darius I put down the revolt and, angered by Athens's backing of the uprising, launched the Persian fleet against Greece. The fleet faltered on the rocks off Mount Athos, and Darius withdrew. In 490 BC, Persia launched a second attack. The Persian army massed on the plain of Marathon outside of Athens where the Athenian army, outnumbered three-to-one by the invaders, won an overwhelming victory. Not about to give up, the Persians, in 480 BC under Darius's successor Xerxes, launched a third expedition. The Greeks made a first stand at Thermopylae where, unable to stop the vast Persian army, the Spartan leader Leonidas and more than 300 Spartans and 700 Thespians fought to the death. The Persians burned Athens only to have the war turn against them at sea.

The Persian fleet chased the Greek fleet to Salamis, an island in the present-day Gulf of Saronikós. Under the Athenian General Themistocles, the Greeks feigned retreat and lured the Persian fleet into a narrow straight. Unable to maneuver, the Persian vessels confronted a devastating naval onslaught and, faced with heavy losses and no recourse, scattered to open water. A year later, Greek forces overwhelmed Persian forces at Plataea and drove out the invaders.

The unflinching leadership of Athens during the Persian wars and the dominance of its fleet gave inspiration to the city's golden age of art, literature, and government. Pericles, who became head of state in approximately 460 BC, added new democratic ideals to the constitution that among many reforms paid for jury service and made it possible for the poorest citizen to participate. He also vowed to make Athens the most beautiful and culturally active city in the world. The Athenians built theaters, libraries, monumental gates, and the Parthenon, the great temple devoted to the Goddess Athena. The ideas of Plato and of his student Socrates stirred the Greek mind. The plays of Aeschylus, Sophocles, Euripides, and Aristophanes touched the Greek heart.

Unlike Athens, which emerged from the Persian wars bent to create the future, Sparta emerged as the military-collective it had always been. Clash was inevitable and resulted in the almost three decades of the Peloponnesian War. Sparta emerged the victor over Athens but would not maintain its preeminence. In 403 BC, little more than a year after the conquest of their city, the Athenians revolted against Spartan occupation, restored democracy and independence, and allied with Thebes and Corinth. To counter, Sparta allied with Persia but met defeat against the Thebans in the Battle of Leuctra in 371 BC. This established Thebes as the dominant power in Greece, which angered Athens who turned around and allied with Sparta. While the great city-states were preoccupied with war and changing allegiances, a once insignificant part of Ancient Greece rose to dominant all.

King Phillip II of Macedonia annexed colonies on the coast of Thrace and Macedonia and in 328 BC won the decisive battle of Chaironea against Athens and Thebes. He formed a congress of Greek states; and, at his behest, they appointed him commander of all Greek forces. Phillip declared war on Persia, only to, before he could march against Greece's once again enemy, die at the hands of an assassin. At the time, Athens' leading philosopher was Plato, Socrates convicted of "corrupting the youth" with his teachings and forced to commit suicide by drinking poisonous hemlock. Among Plato's pupils was the Macedonian court physician's son, his name *Aristotle*. Among Aristotle's pupils was King Phillip's son—Alexander.

At the age of twenty, Alexander, to be known as *Alexander the Great*, transformed the Macedonian army into the most powerful in the region. He leveled Thebes and with the surrender of Athens united Greece. He set his sights outward and conquered Anatolia, Syria, Egypt, and Babylon. He reduced the Persian capital, Persepolis, to rubble and driven to rule the known world pushed into India. Alexander defeated a massive Indian army mounted on elephants only to—when his soldiers, fearful at what awaited them in the uncharted regions beyond, balked at being so far from home—turn back and establish Baghdad as the center of his vast Greek empire. Struck down by what was probably malaria, Alexander died at the age of thirty-three. In his short life, he had made Greece the world's greatest power. The leading edge of civilization's advance, however, was poised to shift to another part of the region. Across the Adriatic, an empire that would dwarf Ancient Greece in domain and ambition was on the ascent—*Ancient Rome.*

13

Economics

ALL MODERN SYSTEMS OF economic practice are based on the assumption that resources are by nature limited and human material wants are by nature unlimited. Although the supposition of scarcity has at least to this point withstood time, the way in which we've accommodated it has changed. Over the millennia and in particular over the last several centuries, we've invented a number of economic theories and philosophies and a number of economic models and systems to implement our theories and philosophies. To better understand the economic systems that exist today, a brief look at the evolution of economic theory and practice is in order.

During the Egyptian, Greek, and Roman empires, systems with strict central control dominated economic life. A person in ancient Egypt saw his or her activities under the harsh management of a ruling class, the leaders of which often had the status of gods. In Ancient Greece, rulers maintained a high degree of control but to a greater extent used their power for the good of their people. In the Roman Empire, life was dominated by great estates called villas. On these estates, landlords supervised the cultivation of their land by slaves and former slaves. Workers and their heirs were required by imperial edict to remain on their lands and to serve their lords. As in Greek and Egyptian times, the power of the Roman Empire and its lords was maintained through a strong military class.

During the fifth century, the Roman Catholic Church and the aristocracy in the parts of Europe that would become England, Germany, and Scandinavia introduced a system of central control that dominated economic life in the West for almost a thousand years—*seignoralism*. In this system, a lord, or *seigneur*, headed an agricultural unit of production and consumption, or a farm that could meet the needs of its workers and overseers with limited trade and outside interaction. Through the power of an armed enforcing body, the seigneur judged, punished, and directed the actions of those under his jurisdiction, the *peasantry*, who worked the land.

By the thirteenth century, the rights of the peasants and the responsibilities of the seigneur were clearly defined. The peasants cultivated and harvested the lord's land, but were allowed to farm and even own some land to support themselves and their families. The peasants also had grazing rights and the rights to fuel and building materials, though not often the right to hunt. The lord had the right to tax his people, to take an inheritance tax at their deaths, and to reclaim their lands if they died without heirs. The seigneur also charged payment for the use of the lord's grain mill, bread oven, and other communal property. In some instances, seigneurs went so far as to charge payment for a peasant father to offer his daughters in marriage.

By the end of the middle ages, the rigid control of earlier periods had begun to break down. Peasants sold excess products to buy freedoms from lords, and lords replaced peasants, whom they had a lifelong commitment to care for, with cheaper wage laborers. By the late 1600s, seignoralism had largely disappeared, the deathblow dealt by the chill climate of the little ice age, waves of bubonic plague that periodically swept Europe, and the social upheaval and labor shortages these created. The seigneurs, though, remained socially dominant.

By the mid 1700s, the industrial revolution had taken hold. Rural communities broke apart, and people moved into cities and labored in factories. Subsistence no longer meant farming. It meant working for the wage needed to purchase the food and other items necessary to subsist. Though, in one form or another, money had been in use for millennia, to an extent never before experienced we embraced a monetary economy. Humanity had established the economic framework that would directly lead to the complex and interconnected economies of today.

As our economic practices evolved, our economic theories also evolved. In Ancient Greece, Plato and Aristotle wrote about problems of trade and wealth. Modern economic theory takes us to the sixteenth century and to a form of economic activity called *mercantilism*. The objective of the mercantile economy was to increase the power and wealth of the nation, measured by a nation's stores of gold and silver. This inspired policies to keep wages low and the population growing. A large, poorly paid population could produce more and cheaper trade goods and thus bring into the nation—or into the hands of its ruling elite—more gold and silver.

In the eighteenth century, a school of thought called *physiocracy* influenced economic ideas. The physiocratic doctrine embraced free trade and the supremacy of natural law, order, and wealth. It also inspired the first serious attempt to study the nature of economic behavior. This was conducted by the British economist and philosopher Adam Smith; who, in his work *An Inquiry into the Nature and Causes of the Wealth of Nations*, founded the field of political economics and put forth many ideas that influence economic thought to this day.

Above all, Smith believed in economic freedom and lack of government control, a *laissez-faire* view of the economy. This view founded the *classical school* of economic thought, furthered by David Ricardo, John Stuart Mill, and Thomas Robert Malthus. Classical economists believed in competition, free markets, private property, and minimal government regulation. If left alone, they felt, the economy would self-regulate to achieve the greatest benefit for everyone.

Opposition to the classical school of economics came from early socialist writers, in particular from the nineteenth century philosopher Karl Marx. An exile from Germany, Marx spent most of his life in London, supported by his friend and collaborator Friedrich Engels. Influenced by the deplorable working and living conditions that existed in Europe at the time, Marx felt that the economy was not capable of regulating itself for the common good. In Marx's view, the economy consisted of two classes of people, the *capitalists*, who controlled the means of production, and the workers, or *proletarians*, who provided the labor. The goal of the capitalists was to use their capital—and the control over labor, material, and production resources it provided—to acquire the greatest amount of additional capital with the least amount of capital outlay.

In his *Communist Manifesto*, published in 1848, Marx concluded that the disparity of wealth between individuals and classes produced by capitalism would lead to its downfall. He saw capitalism as a transitional system that would be replaced by *socialism*. In socialism, a person's economic worth wasn't established based on competition and the market value of labor but according to an individual's need and contribution. Marx, however, felt that the capitalists wouldn't give up their power and wealth without a fight and that socialism would only emerge following a revolution by the proletariat.

After Marx, there emerged the neoclassical school of economic thought. For the most part, neoclassical economists such as Leon Walrus, Karl Menger, and William Stanley Jevons embraced the ideas of the classical school. They believed in low taxes, low public spending, and annually balanced budgets. Disparities in wealth, they felt, had nothing to do with class or opportunity. They were the result of differences in human beings—in talent, energy, ambition, and intelligence.

At about the same time, a development took place in the natural sciences that would have as profound of an influence on economic thought as any work in the field. England in the nineteenth century was an empire in the throes of the industrial revolution. As Marx had observed, life was hard and competitive. Men struggled to feed their families. Companies struggled to dominate their markets. International trade flourished, and nations struggled to extend their political and military influence around the globe. Into this atmosphere of social and economic conflict, the theory of evolution took root and the man most responsible for its development was born. In 1859, Charles Darwin published his *On the Origin of Species*. Increasingly, economics and, in the larger sense, all of human activity were interpreted in a framework of survival of the fittest.[1]

In the early 1900s, Marxist ideas took hold in Russia. And, after the stock market crash of 1929, the United States, Europe, and most of the world plunged into the *Great Depression*. The duration and severity of this downturn perplexed economists and challenged the classical view that a free-market economy would self-regulate to create full employment and improve the general well-being. Faced with a rise of fascism and fearful of a spread of communism, politicians demanded a new economic model.

The British economist Alfred Maynard Keynes provided this model. In his work, *A Treatise on Money*, published in 1930, Keynes sought to explain why an economy operates unevenly, why it has periods of booms and recessions, the ups-and-downs of what economists call the business cycle. In his most significant work, *The General Theory of Employment, Interest, and Money*, published in 1936, he put forth the idea that no self-correcting mechanism existed in the economy that could lift it out of a depression and that the only way to do this was through government intervention. In a recession, Keynes felt, government must encourage private investment and make up for any shortfall in private investment through a

---

1. See chapter 43, *Darwin*.

monetary policy of easy credit and low interests rates and through deficit spending to fund public works projects and to sustain the groups most affected by the economic downturn.

Today, economic thought embraces elements of classical, socialist, Darwinian, and Keynesian philosophy. Some economists feel that more government oversight of the economy and intervention in economic activity is desirable. Others feel that less government involvement is the best approach, a modern play on the relationship between market forces and central control established with the advent of civilization. To predict the effect of economic policy, economists have developed complex mathematical models, called *econometric models*, that through the use of computers and a variety of statistical and other analytical techniques attempt to isolate economic variables and to determine what happens when one or another variable is changed.

This carries our discussion to its conclusion and to a point worthy of our recognition. When we look at our economic theories and practices and how they have changed over time, one observation stands out. Not only have our economic theories and practices advanced, they have advanced in an unmistakable direction. As we evolved to greater individuality and a greater sense of "self," society evolved from more to less collective forms. Corresponding to this ascent, our economic theories and practices evolved from those that embraced less individual freedom and more central control to those that embraced more individual freedom and less central control.

14

Civilization III

By the time of Alexander the Great's death in 323 BC, the *Chou Dynasty* had extended its rule throughout most of northern China, and the first documented New World civilization, the *Olmec*, had risen and fallen in the coastal lowlands of modern-day Mexico.

The Olmec, who date between 1500 BC and 600 BC, were noted sculptors that created works that ranged from tiny jade figures to colossal statues of human heads. Of Olmec subjects, the most famous is the "were-jaguar," a religious figure whose features and body were part jaguar and part human infant. A Neolithic culture, the Olmec wrote in hieroglyphics and traded for obsidian and other basic materials. They worked in mud blocks, built burial mounds, and by the end of their reign had advanced to the point of raising simple earthen pyramids.

In contrast, the Chou emperors in China oversaw one of the ancient world's most culturally sophisticated civilizations. Foremost among its teachers was *Confucius*. Believed born the son of a bureaucrat in the province of Lu, Confucius lived from about 551 BC to about 479 BC. He felt that China had fallen on dark times and could only be saved by the wisdom of antiquity. Above all, Confucius is known for his practical insight and his belief in good conduct and proper social relationships. He encapsulated these beliefs in simple statements: "The superior man is satisfied and composed; the mean man is always full of distress."

Lao-tzu, a contemporary of Confucius who is said to have chastised him for his pride, is believed to have been born in the province of Henan and to have served as a court librarian. Lao-tzu taught the philosophy of the *tao*, or "the way," a system of belief that, in addition to pride, rejected worldly desire and embraced plainness, simplicity, and selflessness.

A follower of Confucian philosophy but influenced by Lao-tzu, *Mencius*, 371 BC through 289 BC, believed that all people are inherently good but that one could be turned to evil by those around him. He spoke about the importance of virtue in a ruler and felt that those with power must exercise it for the benefit of the people.

The teachings of Mencius, Lao-tzu, and Confucius founded *Taoism* and *Confucianism*. They also made their way to India where they intermingled with the teachings of *Hinduism* and of Siddhartha Gautama, the Buddha, who lived from about 563 BC to 438 BC and who established *Buddhism*. In the Buddhist view, the human being is in a cycle of birth, old age, and death that repeats though countless incarnations until he or she achieves nirvana, a state free of greed, hatred, and ignorance. Hinduism dates to about 1500 BC, and has been passed through the ages in a collection of hymns, poems, and ceremonial formulas called the Veda. The Hindus embraced a plethora of deities. Over time, Vishnu, the preserver of the universe, and Shiva, the god of creative and destructive forces, became the most important; and, as the idea of monotheism spread, most came to be seen as under a godhead, typically Brahman.

As remarkable as was civilization in ancient China and the Indus and Ganges valleys, no civilization of the time contributed as much to history as Ancient Rome. According to legend, Rome was founded when Rhea Silvia, daughter of the king of Alba Longa, set her newborn sons Romulus and Remus, fathered by Mars the god of war, adrift in a basket on the Tiber River. The brothers were rescued and nurtured by a she-wolf. As men, they returned to where they were rescued and founded the city of Rome. Legend aside, Rome dates to the twelfth century BC and the conquest of what is today the Italian peninsula by a people from Asia Minor called the *Etruscans*. By the sixth century BC, Etruscan civilization had fallen into decline, and the city-state of Rome, which had been under the rule of despotic kings, overthrew the tyrannical Lucius Tarquinius Superbus and in 509 BC established a republic.

Under the Roman republic, society consisted of a ruling, or *patrician*, class and a common, or *plebian*, class. In accordance with a constitution, government was administered by a senate and in place of a king by two *consuls* appointed for one-year terms. Initially, plebs were admitted to the senate but were ineligible for the consul or other high positions. This led to a struggle between classes, which resulted in the drawing of a legal code, the opening of the consul, and the forming of open magistracies.

It also resulted in the legalization of marriage between plebs and patricians, which gave rise to an aristocracy composed of patricians and wealthy plebian families. This aristocracy led Rome into numerous wars, which in 390 BC resulted in its capture and burning by the *Gauls,* on the move out of what is now France. Rome rebuilt and defeated the *Latins,* whose language would dominate the region until the seventeenth century. Victories over the *Volscians* and the *Hernicans* gave Rome control of the central peninsula, and between 343 BC and 290 BC Rome defeated a coalition of *Gauls, Umbrians,* and *Etruscans* in the north and of *Samnites, Bruttians,* and *Lucanians* in the south to command the length of the peninsula.

Its position as a regional power secure, Rome set its sights on the Mediterranean's greatest sea power—the African city of *Carthage* in present-day Tunisia. The first Punic War, as the conflict is called, lasted between 264 BC and 241 BC and ended with the Roman conquest of Sicily followed by Corsica and Sardinia. Matched by Rome on the sea, Carthage advanced into Spain to secure a foothold from where it could strike Rome by land. The second Punic War began in 218 BC when, in one of history's boldest ventures, the Carthaginian general Hannibal crossed the Alps, came at the empire from the north, and ravaged the Italian peninsula for years—only to be recalled to Carthage to defend the city and its territories against a Roman counterstrike led by Publius Cornelius Scipio Africanus. In 202 BC, Hannibal met defeat at Zama.

Rome rose from the ashes of near destruction as the foremost power on the western Mediterranean. In a series of wars, it defeated Macedonia followed by Corinth, which brought Greece under Roman authority. In the Third Punic War, legions stole through the "impenetrable" walls of Carthage and ended the Carthaginian Empire, whose territory became the Roman province of Africa. Campaigns brought Spain under Roman rule; and on his death Attalus III of Pergamum bequeathed Asia Minor to Rome, which established it as the province of Asia.

With its incorporation of Greece, Rome adopted much of the older civilization's art, religion, and philosophy. Greece's democratic ideals fared less well. In Rome, with the senate and magistracies monopolized by the aristocracy, the general Pompey the Great in 59 BC formed a triumvirate with the patrician Marcus Licinius Crassus and the general Gaius Julius Caesar. Crassus died in 53 BC. Tensions built between the generals; and, in an attempt to limit Caesar's power, Pompey threw his weight behind the senate's order that Caesar disband his legions in Gaul. Caesar returned to Gaul, rallied his troops, marched on and took control of Rome, and defeated Pompey in Spain, Greece, and Egypt. There, in Alexandria,

Caesar fell in love with the beautiful exiled ruler Cleopatra. He anointed Cleopatra Queen of Egypt, brought her to Rome as his mistress, and declared himself dictator for life.

Caesar's life, though, wasn't to last for long as a year later republican assassins ended it. The nobility's attempt to restore the constitution, however, was thwarted by Mark Antony, who after falling in love with Cleopatra and taking her home to Alexandria, returned to form a triumvirate with Marcus Aemilius Lepidus and Caesar's grandnephew, Octavian. With the ouster of Lepidus, the triumvirate disintegrated into a civil war between Antony and Octavian that in 31 BC climaxed in the naval battle of Actium. Fearing defeat, Cleopatra—in command of the Egyptian fleet and better known for her political manipulation than for her military prowess—withdrew, and she and Antony fled to Egypt. As Octavian poised to overrun Alexandra, Antony falsely received word that Cleopatra had been killed and in his grief committed suicide. On learning of Antony's death—and when her efforts to woo Octavian had failed and she faced his plan to exhibit her in Rome as a spoil of Egypt's conquest—Cleopatra took her own life.

With the title of Augustus, Octavian marked the first in a line of Roman monarchs that reached a despotic high with Nero. Centuries of warfare and internal unrest had left Rome a mud-block, wooden-roofed slum of thousands. In the year 64 AD, fire broke out and reduced most of the city to ashes. Nero, who had his wife and mother executed and who was accused of starting the fire, began construction of his vision of Rome—the grand and imposing marble city that would later be home to the Pantheon and Flavian Amphitheater. He financed Rome's rebuilding with high taxes and by plundering the provinces. Revolts broke out in Judea and Britain, and in 68 AD the Praetorian Guards rose against the despot who fled the city and committed suicide.

After Nero came the Flavians rulers: Alba, Otho, and Vitellius. The latter was followed by Vespasian and his sons Titus—during whose reign Mount Vesuvius erupted destroying Pompeii and Herculaneum—and Domitian, so brutal as to be murdered. Domitian's death brought forth the era of the "five good emperors." Based on ability and integrity, each emperor selected and legally adopted his successor. The last, Marcus Aurelius, however, was succeeded by his biological son Lucius Aelius Aurelius Commodus, a tyrant who was also murdered. The reigns that followed culminated in 306 AD with that of Constantine the Great, who in a desperate attempt to hold the empire together proclaimed Christianity its official religion and established a new seat of government at *Byzantium*, which became *Constantinople* and today is Istanbul. After Constantine's death, the provinces faced internal dissent and barbarian invasions. The empire declined and with its fall carried the Western world from *Antiquity* into the *Middle Ages*.

15

Socialism

ALL MODERN SYSTEMS OF economic practice are based on the assumption that resources are by nature limited and human material wants are by nature unlimited. As such, supply and demand are central to economic function. We, however, can balance these variables in different ways. Present-day economics is dominated by two systems of allocation and production: socialism and capitalism. In this chapter, we look at the first, *socialism*.

We begin with a fundamental but often overlooked point. At the heart of all forms of economic practice—modern, primitive, socialistic, and capitalistic—is one factor, human creativity. Every economy exists as a result of our work, our energy, our dreams, our ambition, our skill, our drive, and our knowledge. What other than human creativity can underlie economic activity? Though we often take the point for granted, the human ability to create is at the core of our subsistence, at the heart of all things economic.

Moreover, human creativity takes place within the individual. Corporations don't invent products. Individuals who work for corporations invent products. Government doesn't provide services. Individuals who work for government provide services. Machines may produce cars and circuit boards, but men and women design, build, operate, and maintain the machines. We may align our creative energy with that of others, we may form an agency or a company, but the organizational framework in which we work is sustained by our personal creativity. The more opportunity we as individuals have to creatively express ourselves, and thus to engage in economic activity, the more productive the economy.

The individual's creativity, in turn, is motivated by need. Our motivation may be basic, to provide the necessities of life for our families and ourselves. For most in the industrial world, our subsistence needs are not too difficult to meet, and we perceive other needs as essential to our survival and well-being. Some of us work to accumulate wealth, to see the balance in our bank account go up each month. Some of us work to purchase the trappings that display our wealth: a new car, a bigger house, a higher-definition plasma television set. Some of us work for the praise of our boss and the respect of our peers, to occupy a place in the social structure provided by our employer and to advance in that structure. Some of us are driven by our passion, by the conviction that we're meant to do what we do and that we must direct our creative energy to that end. Moreover, our motivation is rarely singular. At one moment we may be driven to earn a paycheck, at another to furthering our advancement on the job, and at still another to pursuing our calling in life. Human ambition is diverse and interwoven.

No matter our motivation, for us to express our creative energy and contribute to the economy, one other factor must come into play—freedom. To produce, we must have resources and opportunity. We must have the room we need to grow and learn and to express our achievement.

Now, let's see how our assumptions of scarcity and unlimited human want and our notions of freedom, creativity, motivation, and individuality come into play in the socialist scheme of production and allocation.

In its theoretically pure form, socialism is based on the notion that we can best develop and allocate scarce resources by communal regulation of the means to produce and distribute wealth. In the socialist scheme, the individual creates goods and services in accordance with the governmentally determined needs of the community. The individual then turns over the wealth he or she produces to the community, which gives back what it feels the individual needs to subsist and continue to produce. Economic needs are accessed, economic activities are planned, and economic resources are allocated through central control.

One might think that in a planned economy, it would be easy to meet everyone's needs. The government or some regulatory body simply mandates what to produce and how it should be distributed. This hasn't proven to be the case, and for a fundamental reason.

The socialist model assumes that the overriding human motivation is social conformity. In a socialist economy, we exist to fill a role in the state. We survive to occupy a niche in the economic machine of society. As such, the socialist model disregards all other forms of human motivation. We have no subsistence needs. At least in theory, the economy allocates what we require to support ourselves to the extent that we can further contribute to the economy. Moreover, any higher needs are disregarded. Those who administer the economy assume that we're driven by no calling other than to do what is expected of us on the job and thus to fulfill our role in society.

When we as individuals have no outlet to express any drive other than the need we feel for social conformity, our motivation is limited. Correspondingly, our creativity and the economic output that results are limited. We have all met or been that bureaucrat who takes his or her break at the same time every day and who does nothing more than what is in his or her job description. How motivated would we be at work if we were forced to give the government our paycheck and, no matter how much or little effort we put into the job and no matter how much or little money we earned, it gave back only the amount it felt we needed to buy food, pay rent, and catch the bus to work? Socialism disregards our individual need to grow and learn—our drive to build a better life for ourselves and our families, our desire to make our job our own by inventing a better way to do it. The socialist model subverts our individuality to further a presumed collective good. As such, it limits freedom and initiative and restricts the creative activity an economy must have to produce the goods and services its members want and require.

Consequently, and as history has shown, socialist governments are forced to demand social conformity and economic output. The leaders of the Soviet Union imposed strict quotas on the output of farms and factories, and workers and managers who failed to meet these quotas were often dealt with brutally. To maintain the stability and output of the state, Stalin in the Soviet Union executed an estimated forty million dissidents. In China's communist revolution, Mao Zedong executed an estimated sixty million dissidents. Socialism is central control. In an ideologically driven socialist society, rulers go to extreme lengths to maintain that control.

Even in the most oppressive economy, however, the spirit of the individual will come through. What socialist nation exists without a black market? The collective farms of the former Soviet Union were vast, but a disproportionate share of the nation's food was produced on tiny plots workers cultivated for themselves. In the 1980s, China's leaders realized that socialism couldn't survive in an ideologically pure form and faced with social unrest opened their economy to business and investment. As the collapse of the Soviet Union and the decline of Cuba, North Korea, and other predominantly socialist nations make clear, in a socialistic economy, economic activity will invariably wind down and, as China's rulers hope to prevent, in the end collapse.

Today, most of us encounter socialism in the form of entitlements and other government sponsored social services. In this sense, every country employs a blend of socialistic and free-market economic practices. Some nations, such as France and Germany, lean toward a planned economy. Others, such as Taiwan and the United States, lean toward a market-driven economy. To whatever extent we may implement it, socialism can't exist without a degree, or in the minds of most economists a substantial allotment, of capitalism.

16

Civilization IV

As the Roman Empire declined, in Africa the kingdom of Ghana arose from an Iron-Age culture dating back to the fourth century AD to become, with the help of Arab traders who penetrated the Sahara in the eighth century AD, an empire rich from the trade of salt, gold, ivory, and slaves.

In the New World, the city of *Teotihuácan* in the Valley of Mexico ruled history's largest Mesoamerican empire. Along the city's "Avenue of the Dead," the Pyramids of the Sun and Moon rose in tribute to the Gods. These accomplishments are particularly noteworthy since, like the Olmec, the builders of Teotihuácan lacked the copper tools of the Old World pyramid builders. By 750 AD and for unknown reasons, the city had been abandoned.

About the same time, *Mayan* civilization rose in present-day Honduras and Guatemala. In its classical period, which occupied the early centuries AD, the Mayans developed writing, geometry, the concept of zero, and a calendar more accurate than the Julian Calendar of Caesar. The Mayans never built cities as large as Teotihuácan or a true empire but lived in loosely aligned city states defined by temple centers with plazas, monuments, and ball courts. They commemorated their rulers with stone slabs that bore the ruler's image and that in carved hieroglyphics described his dreams and achievements. By the tenth century, the Mayans had abandoned their temple centers, their decline thought to be the result of centuries of war between the various city-states.

In China, the *Ch'in dynasty*, from which we derived the word China, came into rule in the fourth century BC. Followers of the "legalistic" school of thought, Ch'in rulers rejected Confucius and felt that the only way to overcome anarchy in society was with the power of an authoritarian government. Between 221 BC and 209 BC, Shih Huang Ti, the "First

Emperor" expanded China's territories and imposed standard weights and measures and a standard written language. He also completed the Great Wall and to solidify his legalistic ideals ordered nearly all books burned and the hundreds of scholars who protested burned with them. To commemorate himself, Shih Huang Ti ordered the construction of the "Terracotta Warriors," life-sized statues of 2,000 horses and 6,000 riders, the remains of which were only recently discovered.

By the third century AD, opposition to Ch'in repression had weakened the central government; and, in 206 BC, the rebel Liu Pang seized power and established the *Han Dynasty*. The Han era, which lasted to 220 AD, is distinguished by three dynastic periods each of which financed territorial expansion with high taxes and each of which ended in disorder and rebellion. Despite constant upheaval, the Chinese traded with Rome along the "silk road" and invented paper, the sundial, and the water clock. They also established schools that graded their students using written exams and along with other subjects taught Confucianism.

In Western history, the period after the decline of the Roman Empire between the fifth and the fifteenth centuries is called the *Middle Ages*. Historians generally break the Middle Ages into three periods: the *Early*, 400 AD through 900 AD, the *High*, 900 AD through 1200 AD, and the *Late* 1200 AD through 1400 AD. The beliefs, customs, practices, and works of the Middle Ages are referred to as *Medieval*. The Middle Ages was followed by and to a degree overlapped with the *Renaissance*, which ran from the fourteenth to the sixteenth centuries.

The Early Middle Ages, also called the *Dark Ages*, opened with the Byzantine Empire out of Constantinople in control of the Eastern Roman Empire and with incursions by tribal peoples into the Western Roman Empire, which in the absence of Roman authority had fragmented into city-states. To provide economic stability, the church and aristocracy in what would become Europe introduced the social and governmental form based on lords and peasants that we spoke about in previous chapters—seignorialism. The Early Middle Ages also saw the birth of a new religion—Islam. Revealed in a vision of the biblical archangel Gabriel, the tribesman Muhammad spread a monotheistic message of the Last Judgment and social justice called the *Koran*. He also proclaimed himself the last prophet and that his message from God superseded all that came before.[1] In what was once the unassailable Roman Empire, the Early Middle Ages ended with Viking invasions from the north, Muslim invasions from the south, and Magyar invasions from the lower Danube.

In contrast with the turmoil of prior centuries, the High Middle Ages was a period of cultural, artistic, and intellectual achievement. In part, this was made possible by *Feudalism*. As opposed to seignorialism, which is between lords and peasants, feudalism is between nobility. In classic feudalism, lords contractually grant land, or fiefs, to barons who raise the armies needed to secure a territory or to drive off an invader. The High Middle Ages also saw the Pope as the unequivocal head of the Roman Catholic Church, a development that tied the cities and kingdoms of what would become Europe into a cohesive social and economic unit. Papal dominance led to the founding of schools and of the first universities, which offered degrees in law, theology, and medicine. Literature and philosophy flourished with the works of Dante Alighieri and Saint Thomas Aquinas. Architects perfected the Romanesque style, which with French and Spanish influence evolved to become the Gothic style.

Europe's military power under feudalism and its cultural integration under the church had another outcome. They made possible the *Crusades*. Initiated to liberate Jerusalem from Muslim control and in response to the conquest of Syria and Palestine by Islamic Turks, and to procure territory for land-hungry barons, the first Crusade was launched in 1095 when European armies joined with the Byzantine Empire to annihilate the Islamic Turkish army in Anatolia and take Jerusalem. In 1144, Muslim forces struck back with the conquest of Edessa; and, a year later, the papacy launched the Second Crusade. Islamic Turks ambushed the German army in Anatolia. The French Army took heavy casualties in a failed attempt to take Damascus, and Muslim forces out of Egypt retook Jerusalem. The third, fourth, and fifth crusades, and the negotiations that followed, saw Jerusalem repeatedly change hands, to along with crusader cities and castles in Asia Minor, the Middle East, and the Spanish Peninsula fall under Islamic rule.

With the arrival of the late Middle Ages, the trade and cultural interaction that grew out of the crusades brought a willingness to confront papal authority strong enough to initiate a centuries-long battle between state and church. The onset of this struggle was marked by open politi-

---

1. See chapter 41, *Islam*.

cal discourse and by the quest for new avenues of spiritual experience. The spiritual movement centered on the search for a direct experience of God, an act once reserved for kings and popes. In the 1200s, this search for meaning took on a new imperative as thousands of miles away on the Asian plains an illiterate, skin-clad Mongol named Temujin declared himself supreme ruler of his people—*Genghis Khan*. Under Temujin's leadership and that of his successors, the Mongols crossed the Great Wall to conquer China and turned west to conquer Turkestan, Afghanistan, Persia, Asia Minor, Kiev, and Moscow and create an empire that stretched from the China Sea to the Mediterranean. Marco Polo's travels to China and his service to Temujin's grandson, *Kublai Khan*, further led people to challenge political and religious authority. With the Black Death in the 1340s, the medieval search for meaning embraced an apocalyptic vision driven by a fervor for messianic salvation.

The political and spiritual dissolution of the medieval world never saw the return of a messiah but pushed back tradition enough to lift Europe out of the Middle Ages and land it in the Renaissance.

Between the fourteenth and sixteenth centuries, Milan, Venice, and Florence led the world in the arts and philosophies. *Niccòlo Machiavelli* defined a secular history and sought to understand how leadership maintained control over the masses. The *Humanists* saw the individual as more important than God or state and interpreted the classical manuscripts of Plato and Aristotle as other than justifications for Christianity. Under the patronage of Rome's popes, Venice's doges, and Florence's Medici family, artists took their craft to new levels. Lorenzo Ghiberti redefined bronze reliefs. Donatello's freestanding sculptures transformed his art. Filippo Brunelleschi developed linear perspective in painting. Raphael, Giorgione, Correggio, Michelangelo, Leonardo da Vinci, and countless other artists and craftsmen were born of the Italian Renaissance, which spread to embrace nearly all of Europe.

The era also saw advances in metallurgy and construction, the large-scale production of gunpowder, the voyages of Bartholomew Dias around the Cape of Good Hope, of Christopher Columbus to the New World, and of Ferdinand Magellan around the globe. As important, the era marked the development of the theory that a century later would climax in modern science. In the early 1500s, Nicolas Copernicus professed the belief that the planets revolved around the sun, which challenged the biblical notion that the earth was the center of the universe.

Inseparable from this and other advances was the *Inquisition*. The roots of the Inquisition run back to the fourth century and the establishment of Christianity as the official religion of the Roman Empire. In 1252, Pope Innocent IV sanctioned torture to extract the "truth" from those suspected of heresy, and in 1542 Pope Paul III established the Congregation of the Inquisition, or *Roman Inquisition*, to combat the *Protestant Reformation* of Martin Luther and any belief counter to that of the church. The accused faced a jury of laity and clergy and were compelled to answer all charges against them. The testimony of two witnesses was proof of guilt, and sentences were pronounced in a public ceremony with the most severe penalty being life in prison, which the civil authorities interpreted to mean to the time of execution. In 1478, at the request of King Ferdinand V and Queen Isabella I, the papal authority approved the *Spanish Inquisition*. Targeted against Jews, Muslims, and Protestants who couldn't support their claim of Roman Catholic conversion, Spain's Grand Inquisitors executed thousands.

17

Capitalism

At its essence, economic activity is a human endeavor. It's the outcome of our drive and ambition, of our inventiveness and creative expression. This brings us to the economic system that most of us live and work within and that proponents claim offers the individual the freedom to exercise his or her creative power to the fullest extent of his or her abilities—*capitalism*. What is capitalism and how does it work?

In contrast with socialism, the capitalist model is based on the idea that we can best achieve our economic ends by letting our needs directly dictate economic activity—by the market. Whereas in socialism, the individual has no motivation other than social conformity to contribute to the economy, in capitalism the individual has an economically direct motivation to contribute—*profit*, the ability to get more out of an economic activity than we put into it. Driven to turn a profit, we create or identify markets and supply goods and services targeted to those markets. Our needs, choices, and desires drive economic activity, and those that profit from this activity do what they can to address and for their benefit influence our needs, choices, and desires.

This introduces us to the idea of capital. Textbooks define capital as the body of goods and moneys from which we derive wealth, or a bigger body of goods and moneys. But capital means something more—opportunity. In a capitalist economy, control over capital gives the individual the means to raise a family, start a business, earn an education, or in some way transform their motivation, as diverse as it may be expressed, into creative activity and contribute to the economy. The flaw of capitalism, of course, is that not everyone has access to capital and thus to the freedom and opportunity to fully exercise their creative power.

To understand this, we need to peel away the layers of economic theory and look at the fundamental mechanism of capitalism, in particular at something called the *capital cycle*.

Say we have 100 thousand dollars and invest our money by building a house. When finished, our house includes 50 thousand dollars of labor and 50 thousand dollars of land and materials. Now, say, due to favorable market conditions we sell our house for 150 thousand dollars. By virtue of the relationship between supply and demand—or cost and what people are willing to pay—we turn a 50 thousand dollar profit.

But there is more to capitalism than revenues minus expenses. We made money, but where did our profit, the extra fifty thousand dollars, come from? The difference between what we put into the economy and what we got out of the economy came from a transfer of wealth within the economy. We ended up with fifty thousand dollars more capital, and the economy, excluding ourselves, ended up with fifty thousand dollars less capital. Factors such as money supply, inflation, deflation, and the value added by our management and the reshaping of raw materials into a finished product influence this situation, but they don't change the mechanism. We now have more resources, and thus freedom and opportunity to creatively express ourselves and contribute to society and the economy, and everyone else has less.

As this process repeats, capital concentrates in the hands of fewer and fewer individuals. We have the situation Marx grappled with more than a century ago and we see throughout the world today: The rich get richer, and the poor get poorer. Wealth concentration isn't a function of greed and human failing, though greed and human failing may come into play. It's an intrinsic, inescapable characteristic of a profit-driven market economy.

Because of the mathematical certitude with which wealth will concentrate in a profit-driven market economy, a mechanism must exist to redistribute capital. Beneath all its layers of legal, political, accounting, and regulatory complexity, capitalism is nothing more than a scheme to create opportunity and thus economic activity by capital redistribution. Capitalism is a system that over the last several centuries has evolved to cycle wealth from areas of the economy where it has accumulated to areas where it can be put to work—the capital cycle.

Here we must distinguish between capitalistic wealth redistribution and socialistic wealth redistribution. Though our politicians may tell us otherwise, they are not the same thing. Welfare and other entitlements make so little capital available to the average individual that they do almost nothing to create economic opportunity. A person on the dole may have the money to buy groceries but not the money to open a grocery store. Moreover, excessive socialistic wealth redistribution, as when taxes are high, draws funds out of the capital redistribution cycle, diluting capital to the point where it becomes harder to pool enough to invest.[1] Economic activity declines. Unemployment and the demand for social services increases, as does government's call for still higher taxes to pay for them. Social programs keep people afloat, stabilize demand forces, and maintain the social stability necessary for a market economy to function but in a fundamental way do little to fuel economic activity and create wealth.

In present-day capitalism, the most important devices employed to move capital from areas where it has accumulated to areas where it can be put to use are an economy's banking system and its bond, stock, and other exchanges. When we save, invest, and borrow, we're adding to or taking resources out of the capital cycle. The more functional our banks and exchanges, the better the flow of capital, the more opportunity we create and enjoy, and the better the economy works. The less the pie is divided and the more it increases in size.

But the economic pie doesn't always get bigger. Due to a natural reluctance to put investment funds at risk, investment capital isn't available to everyone. A multinational corporation may have no problem raising the money to build a chain of factories or retail stores. A family who wants to start a small business may have to mortgage their home to raise funds. Our tools of capital redistribution aren't always effective and fairly administered. As a result, capital concentrates. Opportunity decreases, and economic activity winds down. In time, the economy will face a period of restructuring and, as Keynes felt and most, but by no means all, economists today agree, the need for government to in some way stimulate investment and by doing so to redistribute capital before things pick up and the process repeats. Whereas if left alone a socialist economy will in the end fail, a capitalist economy will demonstrate the booms-and-busts of what economists call the *business cycle*. Many psychological and other factors influence the business cycle. At its core, though, it's perhaps best characterized as the *opportunity cycle*.

---

1. See chapter 21, *Taxes*.

Faced with the ups-and-downs of the business, or opportunity cycle, economists look for ways to better manage the flow of capital. Classical, or laissez-faire, economists believe this happens best when we don't do anything—that the economy will self-regulate to full employment and optimal production and distribution. In pursing his or her own good, Adam Smith felt, every individual is led as if by an "invisible hand" to achieve the best good for all. Most economists, however, think that the economy requires some oversight, some regulatory management. All modern economies embody capitalistic and socialistic elements. It's debatable as to whether a purely socialistic or a purely capitalistic economy has ever existed, at least for any significant period of time. The reality is that the capital cycle operates within a governmentally administered business environment. Difficulty arises when—through over regulation, under regulation, or regulation that benefits one interest at the expense of another—government fails to structure this business environment in a way that optimizes the flow of capital.

This point carries our discussion beyond capitalism in the theoretical sense to our next chapter on economics, *Today's Economy*. We're all familiar with and caught up in the trends and the ups-and-downs of the contemporary economy. From the standpoint of common sense, what's going on in the economy and can we fix it?

18

Civilization V

WITH EUROPE LOCKED IN the struggle between Martin Luther's Protestant Reformation and the papacy's Roman and Spanish Inquisitions, Western Civilization rose from the Renaissance into the *Modern Period*, the *Age of Enlightenment*.

By the beginning of the Modern Period, which dates to about the year 1600, China had advanced through the *Sui, Tang,* and *Sung* dynasties, the latter ending in the Mongol *Yüan Dynasty*. In the fourteenth century, the vast Mongol Empire disintegrated into warring states; and, in 1368, the Buddhist Monk Chu Yün-Chang led a revolt against the Mongols and established the *Ming Dynasty*. Noted for its silk and porcelain, the Ming, under Zhu Di, also constructed the largest fleet of wooden ships ever assembled. In command of a thousand vessels and thirty thousand sailors, Admiral Zheng He led this armada on seven great voyages that carried the Emperor's influence to the South Seas and across the Indian Ocean to the Red Sea and Persian Gulf.

In the New World, the militaristic Toltec Empire had risen and fallen in Mesoamerica; and, by the fourteenth century, the Aztec Empire had taken root in Central Mexico. At the heart of the Aztec Empire, which incorporated as many as five million souls, was the beautiful stone and adobe city of *Tenochtitlán*, where Mexico City now sprawls. In Tenochtitlán, the Aztecs gathered in marketplaces to trade corn, venison, and jewelry and share music, poetry, and philosophy. The Aztecs also climbed the city's great pyramids to witness acts of brutality as horrendous as any performed in Ancient Rome's amphitheaters. Perched high above the streets and buildings, priests with obsidian knives cut out the still-beating hearts from thousands of human sacrifices. In the Andes of South America, the *Inca Empire*, an agriculturally based theocracy, rose in the Cusco valley in present-day Peru to, by the early 1500s, stretch through parts of Chile, Bolivia, Ecuador, Columbia, and Argentina.

The new and old worlds collided when in 1519 the Spaniard Hernán Cortés invaded Mexico and when, in 1532, his countryman Francisco Pizarro landed on the coast of South America. The ease with which a handful of European ships and soldiers conquered the New World's vast empires has long fascinated historians.

Though in some ways advanced, from a military and technological standpoint, the Inca and Aztec civilizations had yet to reach the level of ancient Sumer. The New World pyramids, for example, rivaled those of Egypt in size but were primarily built using simple earth-fill techniques. Copper and bronze tools and weapons didn't appear in the Americas until the eleventh century and were no match for the gunpowder and iron swords of the Spaniards. Backed by millennia of recorded history, the Europeans also knew how to leverage their adversary's weaknesses. Cortés drew on the Aztec myth of a "white" god to enlist the Aztecs to conquer their own kingdom. Pizarro manipulated the sibling rulers of two major Inca factions into bloody conflict and when they had weakened each other ordered their execution and took control of the empire.

In North America, settlers and explorers encountered stone-age hunting and gathering cultures, most without written language. Even the horse, so prized by Plains tribes, had descended from sixteenth century stock brought by the Spaniards. The conquest of the New World wasn't a feat of triumph or an act of barbarism and Western greed and expansionism. It was an inescapable development in the course of humankind's cultural evolution.

This returns us to Europe and the age of enlightenment. In many ways, the advances in this period rest on the concept of Deism, or the belief that God made the world to function according to rational laws. By studying God's handiwork, people could understand the world and even make it better. Galileo drew on this view to further the concept of empiricism, and by doing so founded science as we know it today. The scientific revolution spread to all areas of thought. Naturalists studied anatomy and classified species. Isaac Newton formulated the classical equations of gravity and motion. Adam Smith founded the field of political economics.

Spurred by these developments and Galileo's trial by the Roman Inquisition in 1633, the Protestant Reformation gained momentum. This led to a shift in power and wealth from feudal nobility and the Roman

Catholic Church to middle class merchants. It also led to feelings of nationalism that in 1688 in England gave rise to the sovereignty of the parliament and that, entwined with the Industrial Revolution, led to the American Revolution, between 1775 and 1783, and the French Revolution, between 1789 and 1799—and the rule of Napoleon I.

During the French Revolution, Napoleon I, or Napoleon Bonaparte, commanded the French army in Italy from where he led victories against Egypt and Austria. After his fleet was destroyed by Britain's Admiral Horatio Nelson in the battle of the Nile in 1798, he returned to France and initiated a coup that ended the revolution and placed him in power.

In 1800, he crossed the Alps to defeat an Austrian force; and, in 1804, after Britain had attacked France at sea, he proclaimed himself emperor. In 1805, Austria and Russia allied with Britain, and Napoleon defeated an Austro-Russian force at Austerlitz. Prussia threw her might behind Russia; and, in 1806, Napoleon destroyed the Prussian army at Jena and Auerstädt and the Russian army at Friedland. In 1807, Tsar Alexander I allied with Napoleon who that same year conquered Portugal and appointed his brother Joseph King of Spain.

By 1810, Napoleon ruled nearly all of Europe and had begun sweeping economic and governmental changes. He abolished the remnants of feudalism and established the *Code Napoléon*. This granted citizens freedom of religion and each state a constitution that provided for a parliament, bill of rights, and male suffrage. With the ambition to one day fund and provide universal public education, Napoleon put schools under central administration and opened higher education to all who qualified. Each state was to sponsor an academy or institute of the arts and sciences, and grant incomes to scholars to conduct their work.

In 1812, Napoleon's alliance with Alexander I disintegrated, and he launched an invasion of Russia that ended in a disastrous retreat from Moscow. Europe united against him; and, in 1814, he abdicated and was exiled to the Mediterranean Island of Elba. In 1815, he escaped and marched on Paris, which ended with his defeat at the Battle of Waterloo. Exiled to Saint Helena, a remote island in the South Atlantic, he remained until his death.

Napoleon's reforms laid the groundwork for the emergence of Europe's modern nation-state. But the emperor's reforms were never fully implemented, and this inspired unrest. Most notable among the uprisings were the European revolutions of 1848. In France, laborers faced with poverty and deplorable working conditions demanded economic reform. Poles, Danes, Italians, Germans, Romanians, Hungarians, and others faced similar economic conditions and demanded suffrage and representative government. Revolts were swiftly put down, but left Europe with the concept of *liberalism* and—after Carl Marx's publication of the *Communist Manifesto*—the concept of socialism.

Liberalism at the time, however, wasn't the same as liberalism today. We equate modern liberalism with social freedom but also with high taxes, social entitlements, and government control—many of the values and approaches associated with traditional socialism. In contrast, the liberal philosophy was founded on the ideal of individual freedom and responsibility made possible by limited government.

By the turn of the century, Darwin had published his *On the Origin of Species*. Alexander Graham Bell had patented the telephone, and Marie Curie had discovered Radium. The British Empire stretched to almost every corner of the globe. France and Belgium controlled much of Africa, and Spain controlled much of South and Central America. In China, the *Manchu Dynasty* had risen to power, and Japan had advanced from its *shogun* period to, under Emperor Mutsuhito, modernize its navy and adopt an American inspired system of universal education. The United States had grown to stretch from the Pacific to the Atlantic, and its Civil War had marked a profound step in civilization's rise. Through Antiquity, the Middle Ages, and the Renaissance, in the New and Old worlds, we had practiced slavery and had fought wars to free ourselves only to turn around and enslave our former masters. Despite some attempts and legal measures to ban slavery that date back centuries, in most of the world the practice was common and considered an acceptable form of social structure well into the nineteenth century. The American Civil War had many causes and objectives, only one of which was slavery. With an intent not seen before, however, the Americans fought to abolish the practice of slavery—the concept of slavery.

In 1905, Albert Einstein published his special theory of relativity. In 1913, Neils Bohr hypothesized his model of atomic structure. In 1914, an assassin gunned down Archduke Francis Ferdinand of Austria-Hungry, and civilization faced the next great turning point in the course of its ascent out of the past—*World War I*.

19

Today's Economy

THE ECONOMIES OF THE United States and the world have long faced the ups-and-downs of the business, or opportunity, cycle. Most analysts, however, feel that the nation and world's most recent economic downturn is different—more sustained, more widespread, more deeply rooted. Our politicians also seem more uncertain about today's economy—less able to grasp the situation, less able to agree on an appropriate economic policy. When we look past the details and nuances that the pundits, politicians, and economists focus on and deal with the economy in terms of the basics of economic function, what do we see? From the standpoint of common sense, what is going on in today's economy and, given the economic difficulties that we face, how would we fix it?

As we would expect, our analysis of today's economy begins with the heart of capitalistic economic function, the capital cycle. No matter how an economic downturn manifests—inflation, deflation, layoffs, shortages—in our scarcity-based, socialist-capitalist economy, problems in the economy stem from a root cause. Economic difficulties result when capital doesn't efficiently flow from areas of the economy where it has accumulated to areas where it can be put to work. The capital cycle, in turn, operates within a governmentally administered economic environment, a legal and regulatory structure. For the economy to function properly, we must structure this regulatory environment to facilitate rather than bottleneck the flow of capital.

In this regard, we can impede the capital cycle at two critical points: the level of the borrower and the level of the lender. From the standpoint of the borrower, when we impede the capital cycle, we, through our regulatory structure, make it hard to open or expand a business or to in a tangible way contribute to the economy—high taxes, excessive environmental regulation, overpriced education and healthcare, policies that encourage businesses to move overseas. From the standpoint of the lender, when we impede the capital cycle, we make it difficult for an investor to receive a reasonable rate of return at a reasonable risk. We create policies that reward risky lending or, conversely, that reward excessively restrictive lending. We create policies that encourage investment in offshore companies and in companies that shuffle money around rather than produce tangible goods and services.

The consequences are twofold: first, we slow down the flow of capital. This results in a downturn in the economy or in a segment of the economy. Capital, however, has to go somewhere. Second, we create a capital bubble. When investors don't have the option to at a reasonable level of risk and return invest in companies that produce tangible goods and services, they buy and sell speculating on price. We make money off of money, gamble that whatever asset or security we're investing in today will, by virtue of the speculation of others, be worth more when we sell it tomorrow. Speculation itself isn't the culprit. When we think gas is going up in price, we speculate, take a chance and fill-up today instead of tomorrow. The problem lies is the circumstances in which we speculate. When we bottleneck the capital cycle, we create a pool of paper capital, artificial wealth. And, like a plugged artery the leads to a fatal aneurism, when the speculation that drove the price of the stocks, currency, real estate, or whatever asset we're buying drives up the price too far above the asset's real worth, the bubble bursts. Many economists, for example, feel that the recent housing bubble won't be fully deflated, and proper real estate valuations reached, until the average wage-earner can afford the price of the average house.

In addition to the legal and regulatory framework in which the private sector and the free market operate, government has made available two immediate, though rather blunt, tools to intervene in the economy, or to manage the flow of the capital cycle: *fiscal* and *monetary* policy.

In the United States, monetary policy is set by a system of private banks established in 1913: the *Federal Reserve*, or the *Fed*. By buying and selling government and as of late nongovernment securities, by changing the *reserve ratio*, or the percentage of deposits banks must keep at a Reserve Bank, and by changing the *discount rate*, or the interest rate charged to banks for borrowing from a Reserve Bank, the Fed regulates the amount of money in the economy. The more money, the easier it is to raise capital. But if there's too much money in the economy, each dollar will be worth less and prices will go up—*inflation*.

Government directly intervenes in the economy with fiscal, or tax and spending, policy. *Demand-side* economists think the best way to grow the economy is to direct tax revenue to low income people, increasing their ability to buy and encouraging businesses to expand and hire to meet demand. *Supply-side* economists think the best way to grow the economy is to make tax revenue available for investment, or to, as did presidents Kennedy, Reagan, and George W. Bush, cut taxes, encouraging businesses to start and create jobs thus giving people more money to spend.

As this debate goes around and around, so do the booms and busts of the business cycle. Though arguably sound in theory, fiscal and monetary policies haven't proven to be particularly effective economic management tools. After the stock market crash of 1929, President Franklin Delano Roosevelt, in his *New Deal*, created thousands of government jobs. Without question, these jobs made life better for those who had them, and the nation is still benefiting from the Hoover Dam and many of the large public works projects of the era, but the economy didn't pull out of the *Great Depression* until the fundamental restructuring brought about by World War II and the relaxation of New Deal wage, price, and other restrictions that followed. Today's global economy makes implementing fiscal and monetary policy particularly difficult. Capital travels freely across borders, and an increasing amount of business is conducted beyond the nation's legal and thus regulatory jurisdiction.

As a result of globalization and of our inability to create a regulatory structure and fiscal and monetary policies that maintain a properly flowing capital cycle, and the economic ups-and-downs that result, the United States in recent decades has developed certain trends within the economy.

Most discernable, our economy has evolved into tiers. A small upper class conducts an increasing amount of economic activity while a shrinking middle class and a growing lower class have less opportunity. This is why one statistic tells us the economy is fundamentally strong and another that family-wage jobs are disappearing—why one politician says the economy is poised for a quick recovery and another that it's headed for the worst downturn since the Great Depression and we should be prepared for years of suffering and hardship. It's a matter of political perception and of how one gathers and interprets the statistics.

As significant, we've diverted investment from companies that produce real goods and services to monetary speculation, and substantial segments of the American manufacturing base have withered into virtual nonexistence. The manufacturing of electronics has largely moved to Asia. The manufacturing of textiles, clothing, appliances, and basic consumer goods has largely moved to low-wage, low-regulation developing nations. We've fallen into an economic model that centers on the financial sector and on financial manipulation—on money games. So corrupted has our management of the capital cycle become that, based on our country's military strength and standing in the world, we borrow overseas money and use it to buy energy and goods produced overseas.

This situation has been made worse by government policies that rather than address the issue of basic capitalistic function, prop up the status quo of an import, consumer-based economy. For decades, the Fed has implemented a policy of dumping money into the economy, one that has added fuel to a long line of speculative bubbles. Recently, the government has dumped trillions into the economy in the form of "bailouts" and "stimulus" plans intended to encourage demand, prop-up the price of houses, and with what amounts to printed money shore up banks and maintain the stability of the capital markets. We've created an economy based on consumption and money-shuffling rather than on the production of goods and services that we can use and export. We've engineered an economy-wide monetary bubble that, like any bubble, must at some point burst—leveraged the entirety of the American economy while decimating our ability to repay our obligations.

Such may characterize the American economy today and for the foreseeable future. How then do we fix the economy? Whatever problems we have or will face in the economy—and whatever fixes, fiscal, and monetary policy we will implement—to improve the economy we must do what the politicians and economists have forgotten—return to the basics of economic function and embrace a common sense approach to economics.

Like any economy, our socialist-capitalist hybrid economy rests on the creativity of the individual. For the individual to express his or her creativity, he or she must have freedom—the opportunity to start a business, earn an education, produce and market an invention, or in any number of other ways contribute to the economy. Banks and exchanges are the mechanisms through which the capital cycle functions, the way we make capital available and thus create the freedom and opportunity the individual needs to express his or her creative activity and play a role in the economy. The government, in turn, maintains the business environment in which the capital cycle operates; it regulates our banks and exchanges. Problems in the economy arise out of—and solutions to these problems are found in—this regulatory structure. The role of government cannot be to bottleneck the capital cycle to serve one or another interest group or political agenda. It must be through appropriate regulation to create the economic environment in which the capital cycle can efficiently function and thus maximize the freedom and creativity, and thus the economic opportunity, of the individual—which maximizes the nation's wealth and economic output.

For government to create a business environment where capital can properly flow, certain commonsense fixes are necessary. One, we must eliminate needless business regulation. Certain health, safety, and environmental controls are of unquestionable value. Many, however, serve political rather than practical ends and do nothing other than increase the difficulty of doing business. Two, we must manage imports and exports. Free trade is of value but must be managed in a way that benefits all parties—you and I—and not just the political groups and corporate concerns that lobby for open borders. Just as a living cell must be contained in a membrane to survive, a nation's economy must have a boundary. We can import and export across this boundary, but our nation's economy must exist as our nation's economy. Three, we must restructure our banks and exchanges to facilitate capital flow and minimize speculation. Capital redistribution is what these institutions exist to do. This is why we created them. We must regulate our banks and exchanges to make funds available to the entrepreneur and small businesses owner and not just to the large corporate concern. Four, we must reevaluate tax, fiscal, and monetary policy. As we ask in our chapter titled *Taxes*, who really pays for our government? Should taxpayers fund a seemingly ever larger government? As did the Hoover Dam and other large public works projects of the 1930s, certain infrastructure investments pay for themselves and are of clear value to the economy. Do the stimulus plans and infrastructure investments of today—"green" energy and the weatherization of public buildings—fit into this category?

We must also address broader issues within society. As we'll discuss in our chapter titled *Education*, to have a viable economy we must have an educated population. A remarkable number of small businesses fail each year because, no matter the enthusiasm and how good the ideas and products may be, owners lack basic business skills. Many can't keep a set of books or read a financial statement. In the larger sense, we must be informed citizens, able to elect the best politicians and demand the most responsible governance. Another issue of concern is healthcare and the increasing percentage of the American economy devoted to the healthcare industry. To some extent, this is justified by an aging population. As we'll discuss in our chapter titled *Healthcare*, however, a large chunk of each healthcare dollar is absorbed by the financial workings of the system. Healthcare costs have a direct impact on the economy. How much of the nation's economic energy can we afford to direct into any one industry? Even more pressing is our nation's energy situation, an issue that we'll devote a number of chapters to exploring, including one titled *Energy Plan for America*. Economic activity may rest on the creativity of the individual, but for the individual to express that creativity—to produce the goods and services we need and demand—it takes energy.

Perhaps at the heart of our economic difficulties is an ideological value that since the 1960s has been driven into the national mindset. It's the notion that economic growth is intrinsically bad. Instilled at our colleges and universities, it's the view that economic prosperity, and in the larger sense industrial progress, is in some way wrong—that humanity prospers at the expense of the environment, that the United States prospers at the expense of the third world. The health, living, and environmental conditions in the United States and developed world are vastly superior to those found in the developing world, and it's our wealth and economic prosperity that has made this possible. Socialism is based on the idea of wealth redistribution—that we must divvy up a limited economic pie. Capitalism is based on the idea that we can create wealth—that we can make a bigger economic pie and raise everyone's standard of living. The prosperity of the United States hasn't been achieved at the expense of the environment or of the third world. The United States economy is vastly larger than that of any other nation. To fix the world economy, we must fix the American economy. Though there are instances of exploitation, we do not rape the earth and our neighbors of resources. The belief that economic growth is in some way incompatible with the environment or with the world community is unsupportable.

Whether scarcity will remain a valid assumption—and socialism, capitalism, and our hybrid socialist-capitalist system will remain viable economic alternatives in the future—is an issue left for the economist and philosopher. To fix today's economy, our leaders must move beyond politics and self-serving agendas and create a regulatory environment that optimizes the flow of capital from areas where it has accumulated to areas where it can be put to work producing real goods and services—things that we can use and export. Though we may be in a sustained period of economic contraction, there will be ups and downs along the way. It's not enough for our politicians to take credit for every economic upswing and blame their predecessors for every economic downswing. The role of government cannot be to micromanage our economy and our individual economic activities. It must be to create the business environment where the creativity of the individual can flourish and our economic activities can unfold. We can divide the economic pie or we can make it bigger, the choice and the responsibility are ours.

20

World War I

Sparked by the assassination of Austrian Archduke Francis Ferdinand in 1914, humanity entered its first great conflict of the twentieth century and its first truly global struggle, *World War I*. The First World War began as a clash between Serbia and Austria-Hungary and grew to involve 32 nations, last more than four years, and mobilize more than 65 million soldiers. An estimated 29 million servicemen were wounded or captured. An estimated 8.5 million servicemen and 10 million civilians died.

Like any conflict, the causes of World War I were rooted in the history that preceded it. After Napoleon's fall in 1815, leaders met at the *Congress of Vienna* to divide the European continent into duchies, kingdoms, and principalities. Invariably, borders were drawn for political rather than practical reasons, which isolated people with similar economic interests and ethnic backgrounds—in particular those of German and Italian descent and French speaking Belgians. Isolation led to nationalism, and nationalism fueled revolution. Belgium won independence in 1830. Italy won independence in 1861, Germany in 1871.

Nationalism also brought imperialism and military expansion. The industrial revolution had fueled an explosion of European manufacturing capability and brought the need for foreign markets and raw materials. Economic expansion took place worldwide but was centered in Africa and its colonies. To protect markets and raw materials, nations built large standing armies and navies and developed plans for mobilization and attack. So as not to be caught alone if war broke out, nations also formed alliances. This led to the rise of two hostile military camps: the alliance of Germany and Austria-Hungary and the alliance of France, Russia, and Great Britain.

With Europe armed and divided into opposing forces with often conflicting interests, the political situation was certain to erupt into conflict. Between 1905 and 1913, war nearly broke out over Moroccan independence from France, Austrian-Hungarian annexation of Bosnia and Herzegovina, and an Italian clash with Turkey, which was then part of the *Ottoman Empire*, over a plan to annex Tripoli. By 1914, all-out war was a flashpoint away.

The Serbian bullet that struck down the Austrian archduke was that flashpoint. Convinced that Russia wouldn't back its ally Serbia, Austria-Hungary declared war on that nation. Instead, Russia mobilized. Germany saw this as a threat and declared war on Russia. France mobilized, and Germany countered with a plan to march against France through Belgium. Belgium refused to allow German soldiers on its soil. Great Britain vowed to enforce Belgium neutrality and declared war on Germany. Italy declared war on Austria-Hungary. Turkey came in against Italy; and, to gain control over German territories in China, Japan came in on the side of Great Britain. The battle lines were drawn. World War I matched the *Allies*: primarily Italy, Japan, France, Russia, Great Britain, and later the United States against the *Central Powers*: primarily Turkey, Bulgaria, Germany, and Austria-Hungary.

On the Western Front, Germany moved into Belgium, drove the Belgium army into retreat, and defeated the French at Charleori and a British expeditionary Force at Mons. France and Britain fell back to the Marne River north of Paris. Germany advanced and crossed. The French General Joseph Jacques Césaire Joffre maneuvered around Paris and attacked the First German army, led by General Alexander von Kluck. Kluck had advanced ahead of the main German force and was without their support. In the battles that followed, Joffre and the French drove back the Germans until, by the end of 1914, both sides had dug-in along a line that ran from the North Sea to Switzerland and initiated a stationary battle of trench warfare that would last nearly three years and elevate armed conflict to a new level of brutality.

On the Eastern Front, Russian armies advanced into East Prussia and the Austrian province of Galicia. Allied victory in Prussia seemed imminent when a German army under General Paul von Hindenburg defeated the Russians in the Battle of Tannenberg. In Austria, the course of the conflict also turned against the Allies. By March of 1915, Russia had advanced through Galicia and positioned its armies to march into

Hungary. In April, a German-Austrian force initiated an offensive in Poland. By September, the Central Powers had forced the Russians out of Poland and Lithuania, and Russia withdrew from Austria.

On the Serbian, Turkish, and Italian fronts, matters also fell in favor of the Central Powers. In 1914, Serbia repelled three Austrian invasions only to, after Bulgaria's declaration of war, be overrun by an Austrian-Bulgarian force. That same year, Turkish and German warships bombarded Russian Black Sea ports, and Russia, France, and Great Britain declared war on Turkey. British naval forces bombed Turkish forts at the Dardanelles, and French, British, and Australian troops landed on Gallipoli. The operations amounted to an abject failure, and the Allies were forced to withdraw. In May of 1915, Italy moved on Austria-Hungary and in an attempt to capture Trieste bogged down in a number of indecisive battles on the Isonzo River.

Germany's success in East Prussia, Galicia, and Poland enabled it to transfer 500 thousand soldiers to the Western Front in an attempt to break the stalemate in the trenches. The Allies countered with an offensive on the Somme River. Both sides gained and lost territory; and, despite Britain's use of the first tanks, neither accomplished anything other than to reestablish the existing line between the North Sea and Switzerland. By the end of 1916, the Allies faced a continued stalemate on the Western Front and losses on the Eastern, Serbian, Turkish, and Italian fronts. Leaders confronted the possibility that France, Belgium, Great Britain, and the other Allied democratic republics could be overrun by and absorbed into the empires and kingdoms of the Central Powers. The United States poised to enter the war.

In the United States, President Woodrow Wilson faced a peace movement that opposed all war and an isolationist movement that felt that the United States had no business entering into a European conflict no matter the cost to liberty and democracy. Since the war's beginning, Wilson had appeased domestic political factions with a policy of negotiated peace between the Allies and the Central Powers. In January of 1917, Germany announced that it would engage in unrestricted submarine warfare against all shipping to Great Britain. This, Central Power gains, and the loss of American lives in the German U-boat sinking of the ocean liner Lusitanian led Wilson to reevaluate his position. On February 3, 1917, he broke diplomatic relations with Germany. Peru, Brazil, Bolivia, and other Latin American nations followed, and on April 6, 1917, the United States declared war on Germany.

Despite United States deployment of what would grow to become the nearly two million strong *American Expeditionary Force* under General John J. Pershing—and Japanese, Australian, and New Zealand victories in China, Samoa, New Guinea, and the Mariana, Marshall, and Caroline Islands—the Allies faced a list of setbacks in 1917. German U-boat activity rose to sink nearly 600 thousand tons of cargo a month. Romania and much of Italy and the Balkans fell under Central Power control. The Turkish Ottoman Empire advanced into Arabia, Palestine, and Mesopotamia; and, as a result of the *Russian Revolution*, Vladimir Lenin and his communist party vied for power in that country, and Russia signed an armistice with Austro-German forces, ending involvement in the war.

Allied losses and defeats continued into early 1918, but with America's entrance into the war and that nation's reserves of men, materials, and manufacturing power, the course of the war by most historical accounts was destined to turn.

In the Middle East, British forces broke Turkish lines and took Gaza and Jerusalem. In Arabia, Colonel T. E. Lawrence, better known as Lawrence of Arabia, led an Arab revolt against the Turks. The French and British went on to take Syria and Lebanon and devastate the Ottoman hold in Mesopotamia. Its future uncertain, Turkey broke with the Central Powers and demobilized.

On the Eastern Front, 700 thousand Greek, French, British, Italian, and Serbian troops struck German, Austrian, and Bulgarian forces in Serbia. Defeated, the Bulgarians drafted an armistice with the Allies, and a liberated Romania reentered the war on the Allied side. The allies continued their advance and shattered the Austrian army in the Battle of Vittorio Veneto. So decisive was the defeat that it precipitated a revolutionary movement. Czechs, Slovaks, and South Slaves set up independent states. The Hungarians followed; and, on November 3, 1918, the Austro-Hungarian government at Vienna concluded an armistice with the Allies. The last Habsburg emperor, Charles I, abdicated, which opened the way for the Austrian Republic.

On the Western Front, Germany had launched a March offensive to force a decision in their favor before American troops could take up position in force. The assault drove forty miles into British defended territory until halted with the help of French reserves. During April, a second German thrust pushed further into France, and in June a German attack drove the line across the Marne River. French troops and the American Second Division halted the assault outside of Paris and launched a counterattack that forced the Germans to pull back.

The Allies took the initiative. In August and September, British and French forces drove the Germans over what had come to be called the Hindenburg line. In October and early November, the British moved toward Cambria, and the Americans advanced through the Argonne Forest and broke the German line between Metz and Sedan. The Hindenburg line crumbled, and the Germans fell into retreat. With the outcome of the war certain, unrest in Germany grew. The German fleet mutinied. An uprising dethroned the King of Bavaria, and Emperor William II abdicated and fled to the Netherlands. On November 9, 1918, citizens proclaimed a German republic. On November 11, Germany signed an armistice at Compiégne that ended conflict. World War I officially ended on January 18, 1919, with the *Treaty of Versailles*.

Unlike any struggle before it, World War I had changed the human experience. It brought the first extensive use of tanks and mechanized warfare. It brought the first extensive use of aircraft and submarines. It brought the first extensive use of mustard gas, chlorine gas, and other chemical weapons. World War I led to the *League of Nations*, which would become the *United Nations*. Above all, the war changed the way we related to the world. Never again could a people see themselves as isolated by oceans and mountains. As nations and individuals, World War I invoked within us a sense of who and where we were in the geopolitical sphere. World War I also set the stage for a vastly greater conflict between nations—*World War II*.

21

Taxes

WHEN IT COMES TO understanding the blend of government control and free market activity that defines today's socialist-capitalist economy, no other issue is more misunderstood and as such sheds more light on the ins-and-outs of economic function than taxation. Almost no one on the paying end of the tax equation likes taxes. Almost no one on the collecting end of the tax equation dislikes taxes. Beyond our personal situation and its costs and benefits, what is taxation all about?

As we discussed in earlier chapters, a scarcity-based economic system is a way to regulate human behavior. It's a set of rules and guidelines that—based on the supposition that resources are limited and human material wants are unlimited—assigns tasks and responsibilities and establishes how the material resources brought into existence by these tasks and responsibilities are divvied out. A scarcity-based economic system is an agreed upon way of conducting economic activity based on a mutually accepted view of the world.

To some extent this regulation of economic activity takes care of itself through market forces. As history and archeology have shown, however, to maintain the integrity of the community in a scarcity-based economic environment some degree of central control is also necessary. Taxation is the tool that government uses to pay for basic services: for defense, for fire and police, for schools and education, for the nation's highways and infrastructure. How could a nation exist without a minimal level of administrative bureaucracy? How stable would the United States and, arguably, the world be if the citizens of that country chose not to fund its global military machine? Taxation is also the tool government uses to shuffle wealth in a way that in theory helps to maintain the social cohesiveness of society. If, in today's scarcity-based economic environment, the economy didn't provide those in need with the minimal resources to live, and the government didn't provide those resources and hopefully the means to move beyond such subsidies, what would prevent anarchy and revolution?

In this regard, taxation to fund government services is a form of socialistic wealth redistribution. Based on what in practice is a political assessment, and through the use of the government bureaucracy and the police power of the state, socialistic wealth redistribution takes money from one group of individuals, deemed to have more than they need, and gives it to a another group of individuals, deemed to have less than they need. To paraphrase former president Ronald Reagan, every tax dollar the government gives someone it takes from someone else.

This brings us to the issue of who actually pays taxes and funds the government. The answer to this question may be different than we might think and our politicians have led us to believe. No taxes are or have ever been paid by a corporation or other legally constructed business entity. All taxes are paid by the individual.

Routinely, politicians call for higher business and corporate taxes and to shift the tax burden from us, the voter, to the company. Sounds reasonable, until we get over the elation we feel that some faceless company will be saddled with the burden of financing the government and look at it more closely. The money to pay a corporate or business tax comes from two principle sources: the customer and the owners or shareholders. A business or corporation isn't a tangible entity; it's a legal entity. To pay a corporate or business tax, a company must incorporate the tax into the cost of its products or reduce the profit of its owners. You and I pay a business tax at the time of purchase, or the owner pays it in the form of a lower return on investment. And most often the owner is not the greedy billionaire whom we so often demonize. Most partnerships and single proprietorships are small businesses—owners often struggling to keep them afloat. With regard to a publicly held corporation, we are the owners, if not directly by way of our pension and mutual funds. To take it a step further, where does the money to post a profit come from? Ultimately, it can only come from one source, the goods and services a company manufactures and sells.

Take the all-too-often, and sometimes justifiably, maligned oil industry. Recently, gasoline prices and oil company profits were, in overall dollars, their highest in history. In response, politicians called for a *windfall profit tax* on oil companies. In a scarcity-based economic environment, where management is charged with the task to maximize profit and return on investment, a business tax is and can only be embedded in the cost of the product. If as a consequence of market forces and sound management, profit and owner return are maximized, where else could the money come from? A business tax is a cost of doing business. Incorporated into the purchase price, you and I pay it at the pump or cash register. Politicians want us to believe that a windfall profit tax on big oil will take money from the rich and give it to the people. We all love to hate the oil companies; and, due to the vital service they provide, they must be held to a high degree of accountability. But unless the government also fixes prices, a windfall profit tax on oil does nothing more than take money from the people and give it to the government. Something the politicians who call for such a tax don't tell us; but, we can be assured, they're well aware of.

An economy consists of us, of the men and women who participate in the economy. Sales taxes, income taxes, business taxes, and fees for government services may favor one group or jurisdiction over another but in the end break down to different ways to collect the taxes that you and I pay. As we discussed in our previous chapters on economics, every form of economic activity is fundamentally an outcome of our effort as individuals—of our dreams, ambition, and creative energy. Consequently, all taxes are, and can only be, paid by the individual. A tax is a burden imposed on the individual's creative efforts, on you and I—hence the name "tax."

This brings us to another misconception about taxes, this one about how a state, federal, or other government budget works. Most of us see the economy as static. We think of the federal or other government budget as a checkbook. If we raise tax rates, the government has more money to spend. If we lower tax rates, the government has less money to spend. But the economy isn't static; it functions through the flow of capital—the capital cycle. Taxes are collected on economic activity: on payment of a salary, purchase of a product, assessment of a property value. As such, the greater the number of transactions and the greater their dollar amount, the more tax revenue collected. When the economy grows and more money changes hands, tax revenues go up. When the economy shrinks and less money changes hands, tax revenues go down.

If we raise tax rates too high and draw too much money out of the capital cycle, we limit investment. Economic activity slows down, and tax revenue decreases. In our hybrid socialist-capitalist economy, the private sector pays for the public sector; business funds government. Or, as Great

Britain's Prime Minister Margret Thatcher put it: Socialism fails because we eventually run out of someone else's money to spend.

Conversely, if we lower tax rates and increase the amount of money in the capital cycle, economic activity will pick up and, at least to a point, tax revenue will increase. It's widely accepted that President Kennedy's tax rate cuts in the 1960s, which dropped the top tax rate from in the low 90 percents to just above 70 percent, sustained the post World War II economic boom, which had begun to decline, into the early 1970s. Similarly, tax revenues by some assessments nearly doubled following President Reagan's tax rate cuts in the 1980s, which dropped the top tax rate from the 70 percents to less than 30 percent. In the Reagan years, the often criticized budget deficit would have been far worse had tax cuts not gone into effect. Many feel that George W. Bush's tax cuts also boosted the economy, though it's harder to say because they were comparatively small. Government revenues in the years immediately after the Bush tax cuts, however, exceeded the amount that many critics had predicted.

Arguments on tax policy and maximizing government revenue center on assumptions about the multiplier effect of tax dollars fed back into the economy with respect to the multiplier effect of those dollars if they had been allowed to remain in the economy. In general, economists feel that to maximize tax receipts there is an optimal point of taxation, and any deviation—either by increasing or decreasing taxes—will decrease government revenue.

On the expense side, balancing the federal budget is also not the same as balancing our checkbook. Some government expenses drain the coffers. Others boost the economy and increase tax revenues. Do we need to subsidize the relocation of manufacturing firms oversees? Do we need to fund a study on the racial or sexual stereotypes of college-age males living in fraternity houses, an example that typifies much of government funded research in the humanities and social sciences? Do we need to subsidize the production of corn-based ethanol, a move that has fueled a massive increase in the cost of grain and that in many parts of the world has contributed to hunger and spurred riots and political instability? Often to the point of absurdity, much of what government spends provides no tangible benefit to society or the economy. It merely takes wealth from one group of people and gives it to another. In contrast, the massive irrigation, hydroelectric, and other public works projects of the 1930s, World War II, the GI Bill of Rights or Servicemen's Readjustment

Act of 1944, the Marshal Plan to rebuild Europe of 1948, and the interstate highway system of the 1950s and 1960s have fueled growth and paid for themselves countless times over. The government maximizes revenues when the economy does best. From a strictly economic standpoint, sound public policy directs government resources to activities that offset the drag of taxation.

In a scarcity-based economic system, taxes are collected from individuals and are used to fund government services and administration. To a degree, social and other programs are necessary to meet essential needs and maintain a nation's cohesiveness as a society. If well directed, they can also help to establish the infrastructure needed to catalyze economic growth. Politicians on either side of the tax argument, however, reduce tax issues to sound bites: fairness, equality, for the children, tax cuts for the rich, government is bad. The economics of taxation is complex, and our leaders may or may not understand its intricacies. When faced with a tax issue, consider how a tax factors into the economy as a whole. Measure a tax's value against its real rather than claimed benefits.

22

World War II

So vast and remorseless was World War I that at its end it was said to be the conflict that would end all conflicts. Few understood that it would leave behind it the conditions needed to incite a more vast and remorseless struggle—World War II. In terms of lives, suffering, and money, World War II is thought to have been more costly than the combined total of all wars fought before it. The Second World War involved 61 nations and 1.7 billion people, three-quarters of the world's population. An estimated 110 million servicemen would be mobilized, and an estimated 25 million servicemen and 30 million civilians would be killed. Eleven million Jews and others would be exterminated in Nazi concentration camps, and untold millions would die in Japanese prisons, reprisals, and weapon and medical experiments.

The events that would bring about the Second World War began the day the First World War officially ended, with the 1919 Allied signing of the Treaty of Versailles.

On the close of World War I, France, Great Britain, and the United States emerged as the dominant powers. To prevent Germany from ever again engaging in war, the Allies sought to cripple that nation militarily and economically. They demanded that Germany limit its army and abandon its navy. They gave West Prussia to Poland to form a corridor between Germany and East Prussia. They forced Germany to give up its colonies, merchant ships, and most of its coal reserves. Above all, the Germans—who had surrendered and formed a new government, and who didn't consider themselves more guilty than any other nation for the war—had to accept the responsibility for its cause and pay its total cost.

Faced with the Versailles treaty, the German Social Democratic Party drafted the constitution for a new Reich, the *Weimar Republic*. Confronted with reparation payments and the stigma of military defeat, attempts to overthrow the Weimar government came from the private armies of liberal socialists and conservative militarists. To stabilize this and other uprisings in Europe, President Wilson in the United States led the formation of the *League of Nations*, an international body where countries could air their differences. Disillusioned by Germany's inability to pay its war debt, however, the United States never joined the League of Nations and retreated into isolationism.

Also disillusioned by Germany's inability to pay its war debt, France invaded the Ruhr in 1923 and took control of its coal mines. The Weimar government encouraged workers to passively resist and printed vast sums of money to pay them. The result was economic collapse and out-of-control inflation. So desperate did conditions become in Germany, that in 1924 the international community had no choice but to change its German policy. Foreign loans and reduced reparation payments enabled Germany to reorganize its monetary system and encourage private industry. Economic conditions improved and, in 1926, Germany joined the League of Nations. Then, in 1929, the *Great Depression* struck the world. Millions of Germans faced unemployment and dire economic circumstances. Disillusioned with capitalist democracy, they turned to communism or to another party with the promise of a better future—the Nazi, or *National Socialist German Workers Party,* led by *Adolf Hitler.*

The depression that gripped Europe and the United States was also felt in Japan, but not to the extent as elsewhere. The Treaty of Versailles gave Japan control over German territories in China and the German Mariana, Marshall, and Caroline Islands. A member of the League of Nations, Japan entered into agreements to limit its influence in China and the Pacific. In 1925, a democratic movement led to universal male suffrage and a call to transfer power from the nobility and military to elected cabinets. As in Germany, prosperity and democracy seemed to have taken root in Japan. Then, in 1926, *Emperor Hirohito*, succeeded to the throne and, influenced by the militarist general *Baron Tanaka Giichi*, reconsidered his nation's future. By the time the Great Depression struck, Japan was isolated from its effects by a controlled economy established to implement a policy of militarization and territorial expansion, a policy Hitler in Germany and *Benito Mussolini* in Italy would also adopt.

In 1931, Japan took control of Manchuria. In 1933, Hitler was elected German chancellor and established himself as dictator. In 1936, Spain fell into civil war, and Mussolini seized Ethiopia. In 1937 and 1938, Japan occupied Chinese ports and aligned with Italy and Germany to create the Rome-Berlin-Tokyo Axis. In March, 1938, Hitler annexed Austria. Unable to face another war, the French and British accepted Hitler's claim that Austria was an internal matter. The way open, Hitler threatened to annex the Sudetenland area of Czechoslovakia. British Prime Minister *Neville Chamberlain* initiated the *Munich Pact*, which gave Hitler the Sudetenland in return for his promise not to take more Czech territory. In 1939, Hitler seized the rest of Czechoslovakia and in a blitzkrieg took Poland. Germany and the Soviet Union signed a neutrality pact. France and Britain, who became the initial nations of the *Allies*, declared war on Germany, who with Italy and Japan became the initial nations of the *Axis*.

Militarily, economically, and psychologically, though, the Allies were unprepared for war. Hitler knew this; and, in May of 1940, ordered a German panzer, or armored, division to move into France. There, he faced a French and British army in disarray. The Germans forced the Allies onto a beachhead near Dunkirk, where destroyers and small craft evacuated more than 300 thousand men in one of history's most desperate rescues. By the end of June, Italy had declared war on France and Great Britain, and the French had signed an armistice that gave Hitler control of the Atlantic coast and the northern part of the country. Germany set up a puppet regime in the town of Vichy, the *Vichy* government.

After the fall of France, Germany launched the *Battle of Britain*, an air war against the British Islands to soften its cities and defenses and destroy its air force prior to invasion. In the United States, President *Franklin Delano Roosevelt* appealed to a nation whose politics were driven by an antiwar-isolationist movement and gained the support to began conscription and increase aid to Britain's new prime minister, *Winston Churchill*. Led by Churchill and backed by the United States, Hitler failed to break the British spirit and in September, 1940, postponed his invasion in favor of a policy to blockade the islands and starve Churchill into submission.

In response, Roosevelt authorized convey ships to attack Axis war vessels, which placed the United States in an undeclared war. In the spring of 1941, Hitler overran Yugoslavia, brought Hungary, Romania, and Bulgaria into the Axis, and, led by General *Erwin Rommel*, rolled into North Africa. On June 22, 1941, three million German troops invaded the Soviet Union, a move that negated that country's neutrality pact and aligned Communist Party Leader *Joseph Stalin* with the Allies. By that time in the Pacific, Japan had occupied most of China and moved into Indonesia. On Sunday, December 7, 1941, Japanese planes struck Pearl Harbor. On December 8, the United States declared war on Japan. On December 11, Italy and Germany declared war on the United States.

Through much of 1942, the United States was occupied with building its military capability, and the war fell in favor of the Axis powers. In Europe, Hitler continued his assault on the Soviet Union and despite harsh winters and high fatalities reached Moscow and besieged Leningrad and Stalingrad. In the Pacific, Japan took British Hong Kong, the Gilbert Islands, Guam and Wake Island, Burma, Malaya, Borneo, and Singapore. On April 9, American and Philippine forces surrendered at Bataan, which for those captured began the *Bataan Death March*.

For the Americans, the tide turned in June of 1942 with the *Battle of Midway*. Deciphered Japanese radio messages let *Admiral Chester Nimitz* take by surprise an approaching Japanese force of nine battleships and four carriers under *Admiral Yamamoto*. In one five-minute strike, American dive-bombers destroyed three Japanese carriers, and a fourth went down later that day. The Americans took the offensive, and on August 7, 1942, took Guadalcanal.

By the end of 1942, the tide had also turned elsewhere on the warfront. The Allies had stalled or driven back the Axis powers in North Africa and the Soviet Union. In 1943, Allied forces landed on Sicily, a move that led Italy to sign an armistice. Not long after, partisans shot Mussolini and his mistress while they attempted to flee to Switzerland. Hitler launched another assault on the Soviet Union, but a counterstrike ended the siege of Leningrad and successive Soviet offensives pushed the Germans off Soviet soil. On June 6, 1944, British and United States troops began operation *Overlord*, D-Day, and established beachheads in Normandy. By the end of the month, the Allies had landed 850 thousand men.

Led by such names as Patton, Bradley, Dempsey, De Gaulle, Eisenhower, and Montgomery, the Allies liberated Paris, pushed north across the German border, and with B-17 flying fortresses hit German targets in night and daytime raids. By the end of November, the Soviets had control of Romania, Bulgaria, Finland, Yugoslavia, and Hungary, and the Americans had marched into Rome. In an act of desperation, Hitler launched a failed last offensive to regain his hold on France, the *Battle of the Bulge*. In February, 1945, Stalin, Roosevelt, and Churchill met at *Yalta* to discuss postwar policy. In April, the United States took the Ruhr, liberated the first concentration camps on Nazi soil, and reached the Elbe River. Soviet armies closed from the East. They met American forces, and the powers agreed that they would partition Germany into two parts. On April 30, Hitler committed suicide, and on May 7 German forces submitted to unconditional surrender.

In the Pacific Theater under General MacArthur, the United States, New Zealand, and their allies had taken the Solomon Islands, struck New Guinea, and scored victories in the *Invasion of the Philippines* and the *Battle for Leyte Gulf*. Early in 1945, Japan introduced a tactic of desperation, the *Kamikaze*. In suicide missions, kamikaze pilots flew explosive laden planes into Allied naval and other targets. Despite the Kamikaze threat, the Americans took Saipan and Iwo Jima, which brought airfields in range of Japanese cities, and began routine bombing of Tokyo. In April, the United States landed on Okinawa where after an easy beginning it faced one of the most brutal defenses in the history of warfare. So unwilling were the Japanese to accept defeat that hundreds of soldiers jumped off cliffs to avoid capture.

Roosevelt died in April, and *Harry S. Truman* became the 33rd president. A pragmatic man, he knew that if the Allies were to invade Japan it would defend its homeland to the last breath and allied casualties would number in the hundreds of thousands. An invasion of Japan, though, wasn't Truman's only option.

In a desperate race against Hitler's scientists, the United States had taken the lead in the *Manhattan Project* to build a nuclear bomb. Prior to the end of the war, it had constructed three devices. On July 16, 1945, it exploded the first in a test near Alamogordo, New Mexico. On August 6, it dropped the second on Hiroshima, Japan. When, on August 9, the Japanese refused to concede defeat, it dropped the third on Nagasaki. Japan faced a Soviet invasion of Manchuria and a nuclear armed United States. On August 14, 1945, Japan agreed to the Allied terms of unconditional surrender. World War II was over.

As with the First World War, however, World War II created the conditions that would lead to still more conflict. The Second World War left humanity with the nuclear bomb, Stalin in the Soviet Union, Europe divided into East and West, and an entirely new form of international struggle—the *Cold War*.

23

Politics

EVERY DAY WE SEE and deal with politics. We face politics at home, work, and school. There are local, state, national, and international politics. There are liberal, conservative, economic, environmental, human rights, animal rights, and every other kind of politics. Issues aside, politics, from the word "power," is best understood when we contrast it with what politicians like to think of themselves as offering and what we often mistake politics for—leadership.

As we know from our chapters titled *Family* and *Society*, social structure is the network of relationships we maintain with one another. In such, we find the basis for politics.

In any society, members hold positions. Take the wolf. Based on age, sex, and other characteristics, members of the pack have different responsibilities and relate to one another in different ways with at the top the *alpha-male*. Positions define wolf social life and make it possible for the animals to get along and in the hunt to coordinate their activities. This is why a lone wolf will often starve, and a pack can bring down prey as formidable as a buffalo. Wolf social structure is as definitive of the species as is the animal's canines and sense of smell.

We humans also have different responsibilities and relate to one another in different ways. Parents and children have their place in the family. Families have their place in the community. Cities have their place in the state. States have their place in the nation, and nations have their place in the larger structure of world society. Just as a wolf forced out of the pack will nearly always die, the human being can accomplish little alone. Through our interaction with one another, and through our attempts to understand this interaction, we create art and music, science and religion, products and businesses, engineering and architecture.

Moreover, our relationships with one another change over time. Friendships come and go. Business partnerships are made and broken. In school, we had our peer group and our place in our peer group. As we grew older and our interests changed, our place in our group changed or we shifted groups. Corporations form and dissolve. Government agencies emerge and disband. National interests align only to, in different circumstances, be in opposition. We adjust our social relationships to create a situation in life that meets our needs and ambitions.

Politics is the process through which we perform this adjustment. Consciously or unconsciously, it's the techniques we use to make friends, establish bonds, and reorganize social networks. It's the chess-game of social arrangement, the strategies we employ to modify the relationships between ourselves and those around us. And, just as in the wolf pack, animals bite and growl to determine which one will be the alpha-male, the "professional" politician fights to make it to the top of whatever social-political hierarchy he or she is climbing.

In contrast, there's leadership. Whereas politics has to do with real or perceived social position and the bases of support needed to maintain and advance social position, leadership is about vision.

To be a leader is to have a dream of the future, an image of where we are and of where we want some aspect of the world to be. What kind of a city, state, or nation do we want to live in? Where do we want to be six months from now, a year from now, five years from now, a generation from now? A leader is driven by the need to bring about change—motivated by the cause of transformation.

But a leader must have more than a dream and desire. He or she must have ingenuity. A leader must be able to draft a plan to achieve his or her dream and must have the organizational skills to implement that plan—the character to inspire others to follow. Leadership is vision and command of the social and engineering practicalities needed to bring vision from mind to reality.

A leader must also be able to look squarely at reality. No future will ever turn out quite the way we plan. And it shouldn't. We learn as we go. Time alters our needs and objectives. We evolve as leaders and as societies. A leader must have the ability to face circumstances and in light of circumstances to refine and perfect his or her vision of the future.

Leadership is about reinvention. We may lead in a small way: a team leader, a school leader, a company leader, a community leader. We may lead in a grand way: Driven to create a better world for our children, our vision is to advance the human cause.

But to place ourselves in a social position where we have the power to reinvent our future and inspire others to devote their lives to our vision of a better world, we must engage in politics.

From humble beginnings in Austria, Adolf Hitler's dream of Aryan superiority and world domination appealed to a Germany humiliated by defeat in World War I. As we described in our chapter on the history of World War II, he reorganized the *German Workers Party* into the *Nazi*, or *National Socialist German Worker's Party*. With the Great Depression in 1929, he recast his message of racial hatred and contempt for democracy as a Jewish-Communist plot and—with the promise of jobs, prosperity, and national glory—was elected Chancellor. Once in power, Hitler banned political opposition and ordered thousands imprisoned in concentration camps. In 1936, he sent troops into the Rhineland. In 1938, he occupied Austria. In 1939, he took control of Czechoslovakia and Poland. In 1940, he overran Denmark, Norway, the Netherlands, Belgium, and France.

Across the Atlantic, President Franklin Delano Roosevelt understood the Nazi threat and knew that the only nation large enough to counter it was the United States. But not all shared his view. Faced with a powerful isolationist movement and a call by pacifists against all forms of war, Roosevelt fought to enact legislation in support of France and Great Britain. In 1940, he secured passage of the first peacetime military draft. In 1941, he drove The Lend-Lease Act through Congress, authorizing arms shipments to Churchill in Britain and to Stalin in the Soviet Union. But it wasn't until the bombing of Pearl Harbor in December of that year that the United States declared war on Japan and entered the struggle against Italy and Germany.

Good and evil aside, Hitler and Roosevelt were among history's most noted leaders. With skill and instinct, they used politics to implement their conflicting visions of the world.

Politics is the tactics and strategy of social rearrangement. Leadership is a vision of the future and the ingenuity to transform that vision into reality. Politicians seek power for the sake of power. Leaders use politics to attain the power they need to implement their idea of change. On occasion, however, circumstances transform a politician into a leader.

24

Cold War

THE COLD WAR MEANS different things to different people. To some it was a battle between military superpowers: the United States and the Soviet Union. To others, it was a battle between economic ideologies: capitalism and communism. In its purest sense, the Cold War was a battle between the past and the future. As we will discuss in our chapter titled *Democracy*, government has evolved from forms with a great deal of central control and little individual freedom to forms with less central control and a great deal of individual freedom. Beneath the Marxist propaganda of a "progressive system for the people," the Soviet Union at the end of World War II was the embodiment of the rigidly governed empires of old.

Properly, the Cold War began at the close of World War II, which left the Soviets in control of Eastern Europe, the Americans and the Allies in control of Western Europe, and Germany, including the city of Berlin, divided into East and West sections. This situation incited a series of moves and countermoves between the Soviet Union and the United States that led to a standoff with dire consequences.

Stalin failed to hold free elections in Eastern Europe. Truman refused to send reparation payments to the Soviet Union. Stalin sought influence in Iran and Turkey and declared ideological war against the United States. The *Truman Doctrine* backed anticommunist forces in Greece and Turkey and sought public support for what in 1947 the journalist Walter Lippmann coined the "Cold War." In 1948, Stalin blocked land access to Berlin, which though under the West's jurisdiction was on the Soviet side of the east-west line. This sparked an eleven-month airlift of food and other supplies into the city and an anticommunist movement in the United States given voice by Senator *Joseph McCarthy*. It also helped to push through Congress the *Marshall Plan* to rebuild Western

Europe. In response to these and other measures, Stalin increased control in Eastern Europe. Truman countered by creating an independent West German state and by establishing the *North Atlantic Treaty Organization*, or NATO. Stalin countered by creating a Soviet led alliance of communist states, the *Warsaw Pac*.

In 1949, this situation climaxed in a most frightful balance of power. The Soviets exploded their first atomic bomb, a move that precipitated a nuclear arms race with the United States and a standoff dictated by a policy of *mutually assured destruction*. If one nation attacked, nuclear retaliation would assure the other's annihilation.

In 1950, Mao Zedong seized power in China and allied with the USSR. The nuclear arms buildup, however, made it impossible for Mao and Stalin to implement a policy of world communism through direct confrontation with the United States or its allies. China and the Soviet Union sought to expand their influence in a less obvious way, based on the theory that one nation would spread communism to the next—the *Domino Theory*. This policy was put to the test in 1950 when communist North Korea invaded Western South Korea. With the support of the United Nations, Truman responded with American conventional forces. Three years—and, on both sides, an estimated 450 thousand lives later—American backed South Korean forces signed a truce with Soviet and Chinese backed North Korean forces to leave the country divided as it had been, along a line north of the 38th parallel that remains to this day.

After Stalin's death and President *Dwight D. Eisenhower's* election in 1953, the Soviets faced a threat of their own making. Under the Martial Plan and the West's free-market economic policies, Western Europe had rebuilt and prospered. Under the East's centrally controlled economic policies, conditions had stagnated to the point where Eastern Europeans had become desperate and were fleeing to the West. Stalin's successor, Communist Party Leader *Nikita Khrushchev*, responded by reinforcing the barrier that restricted travel and communication between the Soviet Union and its satellite states and the outside world—the *Iron Curtain*. This climaxed in 1961 with the building of the *Berlin Wall* between the city's East and West jurisdictions.

Khrushchev, though, denounced Stalin's brutality and favored the buildup of Soviet nuclear over conventional arms. This policy catalyzed superpower tensions and led to the 1962 *Cuban Missile Crisis*. Events began when Cuba's communist dictator *Fidel Castro*—at odds with

President *John F. Kennedy* over the attempted *Bay of Pigs* invasion by CIA backed Cuban exiles in 1961—agreed to let the Soviets target the United States with nuclear missiles placed on Cuban soil. Kennedy responded with a naval blockade of Cuba and the demand that the USSR remove the weapons. For several days Kennedy and Khrushchev communicated through diplomatic channels, and the superpowers poised to respond to a nuclear attack. In the end, Khrushchev agreed to withdraw the missiles in return for Kennedy's promise not to invade Cuba and to remove missiles in Turkey directed at the Soviet Union.

The Cold War during the Kennedy administration also took a more inspiring turn. In 1957, the Soviets placed the first satellite in earth orbit, Sputnik 1. In 1958, America followed with Explorer 1 and a year later created the *National Aeronautics and Space Administration*, NASA. On April 12, 1961, the Soviets launched the first man into orbit, cosmonaut Yury A. Gagarin. On May 5, 1961, NASA and the United States launched Alan B. Shepard into orbit. That same month, President Kennedy instituted the Apollo program and announced to the nation that "before the decade is out" Americans would land a man on the moon and return him safely to the earth. The Soviet Union and the United States had become locked in an arms race and in a space race.

On earth, though, a World War II era squabble would erupt into a conflict that would involve forty-nine nations—*Vietnam*. After Japan's defeat in the Second World War, a communist guerilla named *Ho Chi Minh* led a struggle for Vietnam's independence from France that left the Soviet and Chinese backed Hanoi government in North Vietnam and the Western and United States backed Saigon government in South Vietnam. To liberate the South from "the ruling yoke of the U.S. imperialists...," the North, or *Vietcong*, initiated a campaign of guerilla warfare. Kennedy responded with eleven thousand troops, considered military advisors, a number that after his 1963 assassination, President *Lyndon B. Johnson* escalated to 200 thousand.

Vietnam was a different war in many ways. Battles had no clear lines or objectives but consisted of guerilla attacks on American installations and an American war of enemy attrition. Buddhists, who had no political ties to the Hanoi government, fought the Christian Saigon government, which was unstable and went through coup after coup. Vietnam was also the first war to be primarily managed by politicians. Presidents, advisors, cabinet members, and others in Washington and the United Nations made decisions once reserved for generals in the field.

The nature of the war and its politics came to a head with the 1968 *Tet Offensive*. By that time, American troops exceeded 500 thousand, and Soviet and Chinese relations had cooled to the point where Mao was North Vietnam's only significant backer. In a desperate attempt to break the West's dominance, Vietcong General Vo Nguyen Giap launched a series of coordinated attacks on urban targets that reached into Saigon and even threatened the American Embassy. The American and South Vietnamese response was brutal. North Vietnam lost twenty-five thousand troops, and American generals saw the opening they needed to bring the war to a swift victory. Then something neither side expected took place. For reasons historians debate, Western journalists reported that the Vietcong offensive, named Tet after the Vietnamese lunar New Year, was an American defeat, arguing that if America couldn't secure Saigon after years of fighting they couldn't do so in the future. At home, the news fueled a growing antiwar movement. By early spring, much of the American public had concluded that the war was unwinnable and that the United States had no choice but to withdraw.

Faced with public opposition and even more restrictive political management, the war deteriorated into a series of gains, losses, cease-fires, and peace negotiations—with bombing and other actions reaching into Laos and Cambodia. By the end of 1973, President Johnson's successor, *Richard M. Nixon*, had pulled out most of America's forces. Through what amounted to a public relations victory in Europe and the United States, the Vietcong had overcome what seemed like certain defeat. On April 30, 1975, they overran Saigon and accepted the Republic of South Vietnam's surrender. Peace activists in Europe and the United States celebrated; they had done something that hadn't been done in modern times, forced politicians to end a war. In Vietnam, Cambodia, and neighboring countries, the communist purges that followed took the lives of an estimated three million men, women, and children.

Caught up in the Watergate Scandal, regarding the break-in by Republican operatives at the Democratic National Committee headquarters in the Watergate Hotel, and faced with impeachment, Nixon resigned in 1974. *Gerald R. Ford* became president, followed by *James E. Carter*. By that time, the post World War II economic boom had exhausted. Made worse by inflation, an Arab oil embargo, a Soviet Invasion of

Afghanistan, booming German and Japanese economies, and the seizure of the American embassy in Tehran with the taking of 103 hostages, the United States faced the most severe social and economic downturn since the Great Depression.

With the national mood at its lowest since the years that lead up to World War II, Americans sought a leader with an optimistic message for the future. They found this leader in America's 40th president, a man whom many in the press and academia considered nothing more than a washed-up Hollywood actor—*Ronald Wilson Reagan*.

The United States since Nixon, and the Soviet Union under Khrushchev's successor *Leonid Brezhnev*, had maintained a policy of Détente, or eased tensions. Despite Soviet invasions of Czechoslovakia in 1968, of Afghanistan in 1979, and the backing of numerous third-world insurgencies, the superpowers maintained diplomatic relations and negotiated arms control treaties. Reagan abandoned Détente and launched into a massive arms buildup.

Central to Reagan's military buildup was the *Strategic Defense Initiative*, or SDI. Coined "Star Wars" by the press, SDI was an antiballistic missile system designed to shield the United States from a Soviet nuclear attack and thus make obsolete the policy of mutually assured destruction. Among other reasons, critics blasted SDI on the grounds that the technology wouldn't work. What mattered to Reagan though was not that Star Wars would work but that the Soviets believed it would work.

By the end of Reagan's first term, the largest tax cuts in history had lifted the United States out of recession and launched the economy into a more than decade long period of expansion. In the Soviet Union, however, the centrally controlled economy of old had exhausted itself. On the verge of economic collapse, the Soviets had no way to pay for an arms buildup or to counter SDI.

In 1982, *Yuri Andropov* succeeded Brezhnev. In 1984, *Konstantin Chernenko* succeeded Andropov. In 1985, *Mikhail Gorbachev* succeeded Chernenko. British Prime *Margaret Thatcher* called Reagan—who thus far had refused to meet with a Soviet leader—with the message that we can do business with this man.

Gorbachav knew that the only way to prevent a Soviet collapse was to restructure his economy, *perestroika*, and embrace an openness with the West, *glasnost*. Most of all, he knew that for communism in the Soviet Union to survive he had to avoid the cost of an arms buildup. This meant treaty and negotiation. In Reykjavik, Iceland, Gorbachev proposed the largest arms reductions of the Cold War. Reagan, though, wanted more than coexistence with the Soviet Union. He rejected Gorbachev's proposal because it forced him to give up the one thing he knew would bankrupt the Soviet Union and bring about the fall of communism, SDI.

Gorbachev responded by giving Reagan SDI and an even more sweeping arms control proposal. Reagan accepted, but from the Soviet standpoint it was too late. The past had caught up with the future. Backed by *Pope John Paul II*, and under the leadership of the trade union Solidarity's Lech Walesa, Poland rose in defiance of Soviet control. Gorbachev granted Estonia, Latvia, Lithuania, and other East European nations independence. In 1990, the Berlin wall came down. In 1991, the Soviet Union voted itself out of existence.

25

Democracy

POLITICS, LEADERSHIP, ECONOMICS, AND social structure are closely related. Also in the mix is government. We all like, dislike, turn to, or turn away from government. When we strip away the layers of its implementation and our preconceptions about its role, what is government? In particular, what is the form of government we call democracy and how well does it work?

In earlier chapters, we described the "self" as evolving to greater substance of character, and the family and society as progressing from more to less collective forms. In response, economic systems evolved to provide the individual with greater freedom and opportunity. With ups-and-downs along the way, government also reflected our rise from collectivity.

As seen in the empires of ancient Sumer, Egypt, Assyria, and Persia, despots with absolute power ruled. Less autocratic forms of government emerged in Greece where in the city-states of Athens and Corinth the rule of law predominated and, to an extent that would have been unimaginable in earlier centuries, officials were responsible to the citizens who chose or supported them.

Open discourse inspired the political theories of Plato and Aristotle. In Aristotle's view, good governments serve the general welfare and bad governments serve the individuals in power. Aristotle distinguished three categories of government: monarchy, or government by a single individual; aristocracy, or government by a few; and democracy, or government by the many. Later philosophers defined additional forms of government: theocracy, or rule by religious leaders; bureaucracy, or rule by administrative officials; oligarchy, or rule by a few in their own interest; and tyranny, or rule by an individual in his or her own interest.

Ancient Rome contributed to the development of government with the principle of *constitutional law*. This established the law of the state as superior to the law of lower governing bodies such as the enactments of a city legislature and served to nullify the conflicting interests of conquests and maintain the cohesiveness of the empire.

As we move forward in time to the Renaissance, a demand for constitutional and representative government challenged the power of kings and monarchs, and the nation as we think of it today emerged. In England, the Glorious Revolution in 1688 established the preeminence of Parliament; and, a century later, the French and the American revolutions gave rise to the modern democratic movement. In the eighteenth and nineteenth centuries, democratic ideals strengthened; and, in the twentieth century, the political power of the masses compelled government to take on the responsibility for education, scientific research, resource conservation, and other services.

As the "self" evolved to greater substance of character, society advanced from more to less collective forms. In response, we reinvented government. We progressed from forms that tolerated little individual freedom and participation to forms that accepted greater individual freedom and participation. Governments that provide the freedom and opportunity for the individual to participate are called democratic. Democracy from the Greek *dēmokratiā*—meaning *dēmos*, or people, and *kratos*, or rule—is not a static form of government defined by the vote or by such devices as a constitution and a bill of rights. Democracy represents a spectrum, a measure of the individual's freedom to participate in government administration. The more collective the society the less the value of the individual and the less democratic the government. The less collective the society the greater the value of the individual and the more democratic the government.

This definition is readily apparent in the complex social and governmental structures of today. In our global community, democracy has become the dominant governing form. Most nations embrace, are on the road to embracing, or claim to be on the road to embracing democratic ideals. Our definition of democracy as a spectrum of the individual's ability to participate in government implies that we can carry out democracy in different ways and to different extents. Democracy in Great Britain and the United States is more open to public input than democracy in India and Mexico, which is more open to public input than democracy in

Russia and Venezuela. The extent to which democracy manifests reflects the evolutionary level of the underlying society.

As does the way in which we implement democracy. In ancient Greece and Rome, individuals spoke their mind in public forums. They were free to voice their opinions, and rulers were free to listen or not. We, today, speak our mind in books, television, and on the Internet, and leaders are free to listen or not. Throughout our recent history, people have expressed their opinions in rallies, marches, and public hearings and by forming political groups to take up their causes. In the traditional sense, we implement democracy through the vote.

In this regard, we use two methods of practice. The first is democracy through *direct vote*. In this approach, citizens vote for or against a measure, and the outcome with the majority wins, however we may define majority. Should the city issue bonds to pay for the construction of a new school, yes or no? Should the state of Oregon allow physician assisted suicide, yes or no? Should the state of Florida increase the property tax exemption, yes or no? By way of the direct vote, the individual voices his or her opinion and imposes his or her view on government.

The direct vote personally involves the citizen, but most of us don't have the time to sort through the issues and make informed choices. For this reason, governments also practice democracy through elected leadership, *representative democracy*. In a representative democracy, or a *republic*, rather than vote for or against something, we vote for an individual whom we entrust to understand the issues and enact legislation in our interest.

To further assure that our elected officials have our interests and the interests of our nation or jurisdiction at heart, we limit their power. In the United States, the President, the Senate, and the House of Representatives share power, and are further checked by the courts. We build levels of authority and responsibility into the legislative and political process— checks and balances.

In this regard, we must remember that the vote, direct or representative, isn't democracy but a tool to implement democracy. As such, it can be misused. A majority opinion, for example, can compel the exile or imprisonment of an out-of-favor ethnic or political minority. Similarly, what would prevent a majority from taxing a minority and voting itself perks from the treasury? To prevent situations such as these, democratic governments must embrace values and ideals that are superior to the mechanisms of democracy. A constitution defines the role, limits, and procedures of government. A bill of rights, which in the United States is contained in the first ten amendments to the Constitution, establishes certain individual privileges as inalienable, as based in *natural law*, or law of God and human nature, as opposed to *positive law*, or law of man.

This brings us to the issue of democracy as we practice it today. Do we appropriately use the democratic mechanism of the vote? Does government in the United States and other developed nations allow the individual to participate in a meaningful way?

Beyond the obvious consideration of a free election, for a democracy to function it must meet certain criteria. First, we and our representatives must as much as possible understand the issues. Second, we and our representatives must look past personal betterment and have the interests of society at heart. And, third, we and our representatives must look beyond daily tasks and issues. We must have a vision of the future—an idea of where we are and want to be. If this or any of our criteria is not met, the population doesn't contribute to government in a meaningful way. We may hold elections and embrace the mechanisms of democracy but our votes and activities don't impact government to the extent our leaders may want us to believe.

In today's world, political factions, corporations in particular but also other groups wield a great deal of power. In Washington DC, for example, one of the most powerful lobbies is the environmental movement, which has successfully stopped the expansion of nuclear energy and offshore drilling. This concentration of power, or governance by special interests, has an unavoidable consequence. In our politically driven world, those with power will invariably choose to exercise their power for the benefit of their agenda and for the detriment of competing agendas.

By virtue of the vote and a government's legal and police institutions, democracy pools the power of the citizen, who has little alone, to mediate the actions of economic and other concerns for the greater good. This places political factions in opposition to government. Thus, to achieve their objectives, factions work to manipulate government.

To do this, they corrupt the criteria a democracy must uphold to function. Capital is power. The tool of corruption is money. Economic, environmental, and other concerns spend vast sums to limit our access to information and, by doing so, to center-stage issues and steer votes in a desired direction. Multinational corporations own or back the world's news organizations. Politicians and the public relations firms they hire boil complex problems down to slogans and promises. Environmental groups manipulate the legal system and hire celebrities to appeal to our sense of what is "good" for the earth with little regard for the complexities of environmental issues. In the end, we vote our pocketbook or based on our emotional response to an issue, and our representatives support the groups and causes that fund their political ambitions. As for the vision that will guide us into the future, in our lifetime how many leaders have risen from the political process?

When we fail to uphold the ideals of democracy, we don't have government where we as individuals contribute in a meaningful way but government where factions battle for power. We may hold elections, but our votes don't impact government in accordance with the democratic ideal.

Is this what we see in democratic nations today? Environmental and animal rights groups block forest and wildlife management. Governments bail out bankrupt corporations but fail to fund basic services. A day doesn't go by without word of a scandal over misuse of government funds or over political ties between business and government: bank fiascos, insurance fiascos, accounting fiascos, mutual-fund fiascos, real estate fiascos. Do we have democracy or do we maintain the facade of democracy? Is democracy as we practice it today a viable concept for the future or, as society continues its rise from collectivity, will we demand a yet more open democratic form?

26

Education

ILLUSTRATIVE OF OUR EARLIER chapters on family, society, politics, economics, and government is the issue of education. Faced with a decline in K-through-12 educational performance, liberals call for more money, and conservatives call for strict testing and standards and for vouchers to subsidize private school tuition. Both approaches have been tried and, for the most part, both have failed or been politically untenable. Today, the average high school graduation rate in the United States hovers around seventy percent, with low income and minority students graduating at the abysmal rate of just over fifty percent. As significant is the declining quality of education our schools are providing. In the minds of many, education in the United States is at a crossroads.

In most nations, we've grown up to believe that education serves lofty goals. We go to school to live up to our potential and to become enlightened leaders and citizens. In some respects, education aspires to these objectives, or at least there are those in education who believe it does and who strive to live up to these ideals in their work. But beyond rhetoric and the standards to which some educators hold themselves, we get to the heart of contemporary educational philosophy. Governments mandate education for two principle reasons: to prepare the individual to take his or her place in society and to prepare the individual to take his or her place in the economy.

For a society to be a society, its members must know how to act and behave. In part, school teaches us to fit in. In some societies, the scope of education in this respect is limited to the basics of conduct, with the majority of social skills taught by parents and family. In other societies—in particular in highly collective societies such as those of Cuba and North Korea—education is dominated by social indoctrination. The goal is to graduate students who act, talk, and think alike and in an acceptable way—unwilling to challenge the government.

On the economic front, schools are factories that spit out the human component of global economics—the labor part of land, labor, and capital. Throughout our school years, the system propels us through standardized subjects and measures our performance against that of our peers and established levels of achievement as it pits us in a competition for the economic doors we're led to believe our success will open. Education is an investment with a future return.

However, as we began the chapter by stating, the economic and other returns provided by today's educational approach have over the last decades declined. The failure of the United States educational system was dramatized in the 1980s sitcom *Married with Children* when, excited, *Kelly* handed her high school diploma to her younger brother *Bud* with the words "Read it to me! Read it to me!" There are, of course, many reasons for educational decline.

One is economics. In the United States, the stress of declining living standards affects life at home and public support for education. Faced with a shrinking tax base, the private sector is unable to fund a school system that—driven by security, insurance, technology, healthcare, social service costs, and in many regions a growing and diversifying student body—demands ever more money. In addition, economic conditions often compel families to move from one part of the country to another—a virtual requirement in the agricultural industry and an increasingly significant factor in the professions. Each state employs its own curriculum, and standards vary wildly among jurisdictions. A student moving to a new state will invariably be ahead of or behind in their grade level. Such inconsistency leads to boredom and confusion and may leave the child turned-off to school.

Another factor for educational failure is the decline of the traditional family. More children are raised in single parent households and by grandparents than at any time in recent memory. As important, parents forced to work long hours and multiple jobs to make ends meet have little time to spend with their children and almost no time to participate in their education. In the schools, this is further complicated by the challenge of educating children who come from drug and alcohol influenced home environments and who may have suffered from drug or alcohol exposure during fetal development. Often without the training or in-class support needed to be successful, teachers are faced with the task to instruct children with learning disabilities and behavioral problems.

Educators also complain that to raise reading, writing, and math test scores, and to receive the funds contingent on these scores, teachers are forced to teach to the test at the expense of a well-rounded grasp of the material. Teachers also feel that standardized tests draw time and money from the arts and humanities.

In part, bureaucratization, changes in the social environment, and family, mobility, economic, and related factors such as legal requirements that force schools to offer multilingual courses account for educational decline. But we've long dealt with these issues and, as this would suggest, there must be more to it.

In the minds of some, the greatest failing of educational systems in the supposedly open society of the United States is the degree to which in recent years the emphasis has shifted from core subjects and enlightened discourse to social indoctrination. As interest groups capture the agendas of local school boards and of state and national policy makers, math, science, literature, and history—topics the individual must know to form his or her own opinions—have to an increasing extent been replaced by politically sanctioned subjects. A class on global-warming may fulfill a math or science requirement, but it's no substitute for classes on physics, calculus, and paleontology. A class on minority studies may fulfill a social science requirement, but it's no substitute for classes on the rise and fall of ancient Greece and Rome. A class on free-form art may fulfill a humanities requirement, but it's no substitute for classes on Middle Age architecture and Renaissance sculpture. On the other end of the political spectrum, many schools face pressure to eliminate the teaching of evolution from the science curriculum. To an increasing extent, educators have chosen to reinforce politically sanctioned values, notions that a student grounded in the core subjects and empowered to think for him or herself may or may not choose to embrace.

Another factor is the student. Recently, one of the authors attended a school talent show where his eleven-year-old nephew and eight-year-old niece performed. There, he reflected on something remarkable. Today's student is not the student he or anyone else was years ago.

Most of the performances were teacher supervised and involved the singing of songs one would expect at a grade school talent show—some the authors sang forty years ago in grade school. But there were two acts where the students went their own way, and they stood out—fast, focused, high energy. Children process information faster than we did as kids. One reason, they grew up on the computer. Their brains developed in an environment where feedback and information are at their fingertips, instantaneous. Another reason is society. The days whisk by. The years whisk by. The pace of life has accelerated. We feel it, and we see it in our children. To teach a class today, the instructor must interact with the student differently than in the past or as the instructor may have been taught in college. To be successful, a class has to be tight and structured and it has to move. If you lose the energy, you lose the class. The teacher must demand more from the student, push their limits. Children today are slackers only when they're bored.

But not all educators take the change in our kids and the pace and energy it demands to be a good thing. Teachers complain of discipline problems and short attention spans and blame parents for a lack of interest and involvement. In response, some administrators increase homework and competitive pressures. Others dummy down expectations—require that classes be taught at the level of the average or below average student. Still others ask that students be prescribed medications to settle them down. Rather than redesign lessons and curriculums to better interact with today's student, many educators struggle to roll back the clock.

Particularly disturbing in this regard is an emphasis on group work and social uniformity at the expense of assertiveness, independent thought, and individual achievement. Inspired by a teamwork trend popularized by business schools during the 1980s—and quickly abandoned by businesses when executives saw how little creativity and initiative comes out of a committee—education clings to the idea. The ability to work with others is essential; so is the ability to work alone. It's the Alexanders, the Caesars, the Newtons, the Napoleons, the Curries, the Fords, the Einsteins, and the MacArthurs who have advanced civilization, not the government task force on this and that. We as individuals must be taught to challenge, to question, to think critically—to think for ourselves.

This brings us to another factor leading to the decline of K-through-12 educational performance, one that may not be openly discussed—teacher training and preparation. While there are many creative, dedicated, and well-educated teachers, there are many who are "none of the above." Recently, one of the authors had a conversation with a colleague who reluctantly stated that one-half of all teachers in the public schools were unprepared to be in the classroom—and she is a professor of education at a noted college. Not only does the burden for low teacher performance fall on the student and ultimately the nation and world, it falls on the more dedicated teacher, who must pick up the slack for low performing colleagues. Responsibility for teacher success and failure lies with the teacher and with our college and university system of teacher education.

Today's teachers are predominantly a product of the public education system. In many instances poorly educated when they graduate from high school, they go to college with the same hopes for a future all college students embrace, only to realize that earning mere "passing grades" in their major will not land them a job or allow them to advance in their field. Faced with this reality, many college students are drawn to the decades old academic truism that "if you can't succeed in your major, you can go into education." In recent years, schools throughout the country have been desperate for warm bodies in the classroom, and states have reduced requirements to make it easier for prospective teachers to graduate, become certified, and fill the demand.

But there is a more significant factor than prospective teacher selection. College and university education programs often fail to bring out the best in their students. No matter a prospective teacher's motivation for entering the profession, most prospective teachers have the ability to become fine educators. Universities, however, are not developing this ability. At the university level, education programs are bogged down in bureaucratization and political correctness, and few education professors have the history and above all the science and economic background to separate fact from agenda. Professors must instill in the prospective teacher the lust for knowledge—the drive to throughout his or her career read, learn, argue, challenge, and look for a better way to do things. Those at the pinnacle of the education profession must nurture and demand the highest level of achievement from their students. Professors of education must set the standard for informed and critical thought.

The teacher and the teacher education system may be at the heart of the problem of declining educational performance. They may also be the solution. As individuals, we as teachers and professors must set our own professional standards—and make them the highest possible. The adage that the future lies in the hands of our children is a cliché for a reason—truth. We as teachers and educators carry the weight of tomorrow, and we must embrace the responsibility for this weight with all its magnitude. As teachers and professors, we must look past the system in which we work—past the bureaucracy, past the political correctness, past the agendas imposed by unions, government, and interest groups. We must take as our own a greater role in the human experience. We are the catalyst for tomorrow. The nation depends on us—on the teacher.

Schools need money, but money isn't the solution to all of their problems. Schools need tests and accountability, but tests and accountability aren't the only solution. Schools need parental involvement, and parents need school choice. But parental involvement and school choice are not the only answers.

When the student arrives at school, he or she must be swept up—thrust from subject to subject. In each subject, the student must be immersed in a brief, structured learning experience. To maintain the student's focus, academic topics need to be interspersed with arts, exercise, independent work, and hands on learning. At the end of the day, the student needs to go home excited, exhausted, ready for a good night's rest—not faced with hours of homework. We need homework when it benefits the student, when it reinforces independent thought and personal achievement. We don't need homework for the sake of homework. We as parents, teachers, and administrators need to reinvent the way we conduct our classes. Teachers must take it on themselves to increase their knowledge of subject content. Persons with degrees and backgrounds in pertinent fields should be embraced by the teaching profession, not shunned. As it stands, a Ph.D. in physics or a CEO of a major corporation can't teach in a public high school because he or she lacks an "education" degree. Government, teachers' unions, and the education industry must place the needs of the student ahead of their own agendas. Our educational system must produce students that are more than a product of interest-group indoctrination, that are more than a component of scarcity-based economics. We must produce students that have the knowledge to critically examine the world and to lead their generation to a better way of life. And we must begin by challenging the teacher to above all be a thinker.

27

Healthcare

Healthcare in the United States is an issue that has been in the spotlight for decades and, given political leadership on both sides of the aisle, will probably be in the spotlight for years to come. To make sense out of the arguments on healthcare reform from the political left and political right that we as voters and consumers will face, and the decisions we will be forced to confront, we need to know how our system evolved, what it consists of, and how the American healthcare model compares with systems around the world.

Prior to World War II, healthcare in the United States was run as any other business. A doctor or hospital was a vendor that like a plumber or carpenter provided a service for a cash fee. And for the most part this system worked. To remain in business, doctors and hospitals could never charge more than what their patients could afford to pay or their patients would go to another doctor or hospital. A doctor's payment may have been in the form of an invitation to Sunday dinner; but, such that they were, healthcare services were available to nearly all in need.

This "laissez-faire" approach to healthcare changed during the post World War II economic boom. With roots traceable to late nineteenth century Europe and more directly to wartime industrialization and Franklyn Delano Roosevelt's New Deal during the Great Depression, unions negotiated with companies to provide employees with health insurance benefits. In theory, health insurance is a simple idea. By periodically collecting a little money from each member of a group over a long period of time and by making this pool of funds available to members when unexpected costs occur, insurance reduces the risk that any one contributor will face a bill he or she couldn't otherwise afford to pay. And for decades, the American system of employer provided health insurance worked brilliantly. As it was implemented however, health insurance had a fundamental flaw.

Insurance eliminated the price control mechanism present in the laissez-faire healthcare model. "What does it matter how much the doctor charges, the insurance will cover it." Human nature what it is, costs increased, insurance companies passed rising prices to the employer, unions negotiated for higher employee benefits, and companies incorporated rising costs into the price of their products. This had two consequences. On one hand, rising prices dumped vast amounts of economic resources into the healthcare industry, and America and the world enjoyed an outpouring of new drugs and procedures—decades of medical advance. Without the American system of employer provided health insurance, the medical profession as we know it today wouldn't exist. Like a pyramid scheme, however, the rising costs associated with our insurance-based healthcare system would catch up with us.

Today, healthcare in America is a paradox. We have the finest hospitals, the best-trained doctors, nearly every drug produced, cutting-edge medical technology, and the latest and most innovative treatments and procedures. Conversely, 46 to 51 million Americans have no health insurance, with at least nine million of these children. For the most part, these aren't the poverty stricken unemployed; they have access to Medicaid programs. The uninsured are the working poor: those with multiple part-time jobs or entry level labor and service positions. Even those with insurance often face prohibitively high deductibles and co-payments. By law, emergency rooms can't turn away people in need. The medical bills that follow, though, may saddle the patient and his or her family with bankruptcy and years of financial hardship. Depending on the statistics, America spends between 13.9 and 16.3 percent of its Gross Domestic Product, or GDP, on healthcare—on a per capita basis twice as much as any other nation. Based on access, affordability, and indicators such as lifespan and infant mortality, however, the World Health Organization ranks the United States 39th in overall healthcare performance—behind Oman, Cyprus, Colombia, Saudi Arabia, the United Arab Emirates, Morocco, Costa Rica, and every industrialized European nation.

By any reasonable assessment, the United States healthcare system presents a challenge for the nation and for the doctor, patient, and insurance company. In its current form, the American system consists of a jumble of government and private healthcare providers paid for by a jumble of government and private health insurance pools.

On the public side, the Medicare program is the government's largest healthcare entitlement. It provides health insurance to people aged sixty-five and over and is funded and administered by the Federal government. According to Treasury Department data for the year 2007, this amount totaled 440 billion dollars. Also an entitlement, Medicaid accommodates families and individuals with low incomes and is administered by the states and funded by the states and federal government. According to Treasury Department data for the year 2007, this amount totaled 181.6 billion dollars. There are also veteran's hospitals and veteran's insurance programs, the Children's Health Insurance Program, and a number of state-funded and administered healthcare plans. On the private side, the American health insurance industry is not a free-market system in the traditional sense but is based on regional monopolies. If you want to purchase an individual policy in central Florida, for example, your option is between two major companies, and as one would expect both have similar prices and offerings. Insurance companies do not freely compete; but, based on an entanglement of state and national regulations, divvy-up the available markets.

In part, this conglomeration of public and private health insurance offerings developed to keep prices in check. Medicaid and Medicare stringently enforce reimbursement rates, and doctors often complain that they aren't high enough to cover costs. In the private sector, the Health Maintenance Act of 1973 opened the way for Health Maintenance Organizations, or HMOs. With the goal to check costs, HMOs place contractual limits on fees and services, and doctors often complain that their treatment decisions are overruled by insurance bureaucrats. In an attempt to keep healthcare costs down and maximize return on investment, health insurance providers may deny patients with preexisting conditions from purchasing insurance, and policies may not cover procedures deemed experimental or with a low chance of success. Patients may also be limited to certain hospitals and must choose physicians from the insurance company's provider list.

In part, America's conglomeration of public and private health insurance offerings developed in response to the lobbying efforts of interest groups. Approximately one-third of America's healthcare cost is associated with the insurance industry. Each state independently regulates insurance providers and mandates what drugs and procedures an insurance plan must offer. Lobbying to include coverage for acupuncture, chiropractic, and mental healthcare, for example, has driven up the cost of many insurance plans. On the national level, the pharmaceutical industry has backed legislation to make illegal the purchase of cheap drugs from Canada and to prevent negotiating with drug companies on a "cost-plus" basis. Price checks legislated away, drug companies can charge as much as they want on patented drugs—profit margins of two, three, even four thousand percent are common.

The United States healthcare system embodies a blend of public and private elements and, like any large, public-private alliance, is a legal, political construct that serves a plethora of interests—chiefly those with the best attorneys and lobbyists. As a consequence, many patients can't afford the most basic health insurance plans, and indigent care costs are passed to the insured further driving up premiums. Out of control costs in the private sector drive up Medicare and Medicaid costs, and American companies—forced to incorporate healthcare costs into their products—have little choice but to cut jobs and slash wages and benefits to compete with foreign firms. Inherent at nearly every level of the American healthcare system is the inducement for costs to spiral upward.

Such that it is, how does the American system compare to systems throughout the world?

In England, the National Health Service covers all medical needs, including dental and vision care at no direct expense to the patient. Prescriptions are about ten dollars and are free for seniors and children. The English system, however, has its drawbacks. Doctor's appointments are assigned based on the urgency of the need, and citizens with noncritical needs may wait up to three weeks to see their doctor. Hospital emergency rooms address critical care issues immediately, but wait-times for noncritical care issues may be several hours. There are also limits on the type of treatment available, and horror stories about patients not able to get costly cancer and other treatments and medications abound. Because of wait-times, waiting lists for procedures, and limits on what procedures may be performed, about ten percent of citizens also buy private insurance. For private patients, wait-times are virtually nonexistent and the quality of care is high. The National Health Service is funded by the taxpayer. In total, England spends about one-third per capita on healthcare as the United States.

The Canadian model is also a taxpayer funded government program. All "necessary care" is covered free of charge. Coverage for additional services such as long-term care, physical therapy, and dental and vision care varies from province to province. Thirty percent of healthcare providers are private enterprises that bill the government for services. The Canadian healthcare system negotiates with drug companies on a cost-plus basis, and prices for generic and patented drugs run between twenty and thirty percent of what Americans pay. Though wait-times in city emergency rooms are similar to those in the United States, services may be limited in remote locations. The Canadian system is also plagued by a scarcity of specialists and diagnostic equipment and by long waiting lists for some elective and even for some non-elective procedures—a situation that has forced patients to cross the border into the United States and pay for care. To increase treatment options, in 2005 the Canadian government passed legislation to allow private health insurance companies to cover services not available under the national plan.

Like Canada and England, France has a system that melds private health insurance with government provided healthcare. Even though the nation spends per capita half of what the United States spends, the World Health Organization ranks it number one in healthcare performance. There are no significant wait-times for hospital stays and doctor appointments. Pharmaceutical costs are minimal, and the system includes coverage for emergency home visits by physicians. The key to the efficiency of the French system is low administrative costs—less than five percent. Unlike patients in Canada and England, patients in France pay out-of-pocket for certain services. Patients may also pay for and later be reimbursed for costs. Many French companies also provide employees with private insurance to cover expenses not picked up by the national plan. On the average, the French citizen pays about seven percent of his or her annual income for healthcare services or taxes directed to healthcare services.

We think of the United States healthcare system as a free market model and the British, French, and Canadian systems as socialized models. Each, however, embodies elements of market-driven and socialized medicine. The more socialized systems do well with access, preventative care, and cost. Like Medicare, the national systems set reimbursement rates. To remain in business, therefore, private insurance companies and healthcare providers must keep prices down. Like any socialized endeavor, however, government managed healthcare may be impersonal and foster little innovation. The national plans also place administrative limits on the availability of certain treatments and medications. With some exceptions, in particularly in France, the European systems heavily rely on American funded technology and drug and other research. In contrast, the American system fosters the entrepreneurial spirit but is expensive, has limited access, and emphasizes costly treatment over inexpensive preventative care. Burdened by increasing costs, the American system has also become plagued by waiting periods, limited physician choice, and insurance company bureaucrats overruling the healthcare choices of physicians.

Whether through taxes or insurance, in the end we pay for our healthcare. What is most striking about the American system is the amount we pay. Plans for reform range across the socialist-capitalist spectrum. On the right, there are those who call for a return to a laissez-faire model where everyone pays out-of-pocket. A growing number of doctors, for example, operate on a cash basis. In the middle, there are those who embrace the status quo. We have the greatest healthcare system in the world, they tell us, and we should leave it alone. On the left, there are those who advocate for a national health insurance program, a single payer model. They point out that a health crisis in France, Canada, or England will never result in a patient's bankruptcy and that no one will be refused treatment.

In America, how much of the GDP can we dump into healthcare? Recent proposals from the political left and from the political right have called for a mandate that requires Americans to buy health insurance. Similarly, the Obama administration recently budgeted in excess of 600 billion as a "down payment" on healthcare reform. Would not such approaches dump even more money into an already bloated and out-of-control system? To avoid the upward spiral of prices, mustn't any reform embody some mechanism of cost containment? If so, how do we implement this mechanism? Do we turn to the market or to the government? As significant, mustn't any reform make healthcare available to all who need it, and mustn't any reform attempt to maintain rather than subvert the creative spirit that has driven the advance of healthcare? Though reform of the American healthcare system may seem like an unsolvable problem one thing is certain: We must look past the attorneys and lobbyists and make decisions directed toward an outcome that passes the test of common sense.

28

Earth

IN THE CHAPTERS THAT follow, we deal with issues of energy, population, and environment. More often than not, political discussions on these matters take place without a basic understanding of the earth and its evolution. In this chapter, we provide the insight to address environmentally related topics with background and perspective.

Geologists divide the earth's past into four major stretches of time: the Precambrian and the Paleozoic, Mesozoic, and Cenozoic, eras. They divide eras into *periods* and periods into *epochs*.

The Precambrian opened with the earth's formation from a cloud of stellar gas and heavy element debris that 4.7 billion years ago drew in on itself to create the sun and planets. But the earth at its formation was nothing like the earth we know today. At its beginning, our planet existed as a super-heated ball of molten iron, silicone, aluminum, and other elements. By about 4.4 billion years ago, it had cooled to the point where the crust had solidified, and an atmosphere of methane, ammonia, nitrogen, hydrogen, water vapor, carbon dioxide, and little gaseous oxygen had accumulated. Then, by about 4.1 billion years ago, water vapor in the atmosphere had condensed to form oceans; and, by about 4 billion years ago, the first single-celled life forms had emerged.

The first cells were simple bacteria and blue-green algae most closely related to the single-celled life forms traditionally classified as *Prokaryotae*. The small amount of oxygen in the early atmosphere suggests that these cells maintained an anaerobic, or non-oxygen based, metabolism in which they derived energy by transforming hydrogen and carbon dioxide into methane. By about 3.5 billion years ago, cells capable of creating carbohydrates from carbon dioxide and water while giving off oxygen had evolved. This led to a buildup of oxygen in the atmosphere and to the formation of an ozone layer. By about 1.4 billion years ago, atmospheric oxygen had reached the level necessary for cells to maintain an aerobic, or oxygen based, metabolism.

The use of oxygen increased the cell's ability to downgrade energy from the glucose molecule, which made a more complex form of the cell possible, *Eukaryote*. Eukaryotic cells have a nucleus that contains most of the cell's DNA and structures called organelles that are enclosed by a membrane and contain their own DNA. By about 800 million years ago, *heterotrophy*, or the feeding by one species of life on another, had become widespread, and the biosphere reflected the complexity of this interaction. By about 750 million years ago, Eukaryotic cells had combined to create a more complex life form—the multicellular organism.

This development thrust the earth into its first great geological era, the Paleozoic, which lasted from 540 million years ago to 245 million years ago and contained the Cambrian, Ordovician, Silurian, Devonian, Carboniferous, and Permian Periods. At the onset of the Paleozoic, the surface of the earth consisted of a vast ocean from which rose various landmasses. Most of these landmasses, or protocontinents, were situated in the tropics and southern hemisphere. Life was confined to the seas, which teamed with worms, sponges, mollusks, and other invertebrate organisms as well as with various types of algae and seaweed.

By about 500 million years ago, the protocontinents that would one day form Europe and North America had come together to create a vast continental area, much of which was submerged beneath a shallow sea. In this and other aquatic regions, corals and clams flourished, as did primitive armored fish and other vertebrates. By about 450 million years ago, plants had colonized land, followed, 50 million years later, by animals. At the same time, rays, sharks, and other fishes roamed the world's oceans. By about 370 million years ago, flying insects had emerged, and forests of ferns and woody plants covered much of the land. Ultimately, these forests would decay to create many of the vast oil and coal reserves found in North America and other parts of the world.

By about 250 million years ago, the earth's landmasses had come together to form a single super-continent called *Pangaea*. Not long after, a large number of species died out in what some paleontologists consider the greatest mass extinction of all time. This extinction marked the spread and evolution of a more advanced class of life. The Paleozoic era had ended, ushering in the *Mesozoic* era, which spanned the period from

245 million years ago to 65 million years ago and is known as the age of the reptile.

During the Mesozoic—its Triassic, Jurassic, and Cretaceous periods familiar to every Spielberg fan—Pangaea split into two continents, the southern called *Gondwanaland*, the northern called *Laurasia*. Gondwanaland then split into what would become India, Africa, Australia, Antarctica, and South America, and Laurasia split into what would become Asia, Europe, and North America. Biologically, the Mesozoic saw the rise of many new life forms, but none that has captured our attention more than the dinosaur.

The earliest dinosaurs rarely exceeded 4.5 meters, or 15 feet, in length. They ran on their hind feet and balanced their bodies against the weight of enormous tails. By 195 million years ago, the various species with which we are most familiar had begun to emerge. These included the armor-plated stegosaurus, the rhino-like triceratops, the massive two-footed carnivore tyrannosaurus rex, and the heavy four-footed vegetarian apatosaurus.

The dinosaur may define the Mesozoic era, but the age also saw the rise of many other species. These included birds and modern varieties of such common plants as oak, maple, holly, beech, poplar, laurel, walnut, magnolia, and sassafras. At the end of the Mesozoic, another mass extinction swept the planet. Among the animals to vanish was the dinosaur. The earth entered its most recent geological era, the *Cenozoic*. It was an age defined by the rise of the mammal and by the mammalian order to which we belong—the primates.

The Cenozoic consists of the Tertiary and Quaternary periods and the more familiar Paleocene, Eocene, Oligocene, Miocene, Pliocene, and, in the Quaternary, Pleistocene, and Recent epochs. At the beginning of the Cenozoic, several major groups dominated the mammalian class. The various species that made-up these groups consisted of relatively small creatures, few that exceeded the size of the modern bear. Most were four-footed, had five toes on each foot, and had slim heads with muzzles.

By about 54 million years ago, mammalian life had seen significant changes. The evolutionary ancestors of the horse, camel, rodent, and rhinoceros had emerged, as had the first aquatic mammals, ancestors of the modern whale. By about 38 million years ago, true carnivores had appeared—animals that resembled modern dogs and cats. By about 26 million years ago, various grazing species had evolved, as had their predators.

By about 12 million years ago, modern versions of many well-known animals inhabited the earth. These included the elephant, mammoth, and mastodon, the latter two now extinct. Wolves had also emerged, as had the fox, lion, puma, skunk, otter, buffalo, antelope, hippopotamus, and the now vanished saber-toothed tiger.

The mammalian class also included the primates. Curiously, the first primates were small, rodentlike creatures that resembled modern moles and shrews. But like all primates, they had refined vision with good depth perception and a comparatively large brain with a fissure between the first and second visual areas.

Around 50 million years ago, the primates branched into two suborders: traditionally classified as *prosimians* and *anthropoids*.[1] The prosimians changed little over time and became the modern loris, lemur, and tarsier. By about 38 million years ago, the anthropoid line itself had branched. One shoot gave rise to the new world monkey. The other gave rise to the old world monkey, gibbon, and orangutan and to a number of early apelike species—one of which led to us.

The earliest creatures generally accepted to be in the human line fall into the extinct genus *Australopithecus*, as opposed to the genus *Homo* to which we belong. Fossil remains of Australopithecus date back to about five million years and have been discovered in a number of sites in eastern and southern Africa. The oldest of these creatures stood and walked upright and had a brain a little larger than that of a modern chimpanzee. It also had a sloping forehead, a bony ridge above the eyes, and no discernible chin.

By about 2 million years ago, Australopithecus had disappeared, replaced by *Homo habilis*. Homo habilis had a larger brain and made the first stone tools. By about 1.7 million years ago, Homo habilis had disappeared—replaced by a creature still closer to ourselves, *Homo erectus*. Homo erectus had a larger brain, a higher forehead, and a less distinct ridge above the eyes. It ranged from Asia to Africa to Europe, made stone

---

1. In the less popularly known taxonomic scheme used by biologists today, the order *Primate* is divided into the suborders *Strepsirrhini*, with includes lemurs and lorises, and *Haplorrhini*, which includes tarsiers, monkeys, and apes, including humans.

tools and hunted game as intimidating as the now extinct cave bear—larger than the modern Kodiak. Homo erectus also cooked its food, for it had learned to harness fire.

Between 300 and 500 thousand years ago, Homo erectus gave rise to what from an anatomical standpoint we can best describe as various archaic forms of our own species, *Homo sapiens*. The best known of these archaic forms is *Homo sapiens neanderthalensis*, or Neanderthal. Anatomically, Neanderthal was as tall as we are but more robustly built, with a receding chin and a heavy brow ridge. During much of Neanderthal's 200 to 400 thousand year reign, glaciers periodically swept the northern latitudes. In this environment, Neanderthal bands scavenged mammoth and hunted reindeer, cave bear, and woolly rhinoceros.

At a point in time taken to be about 100 thousand years ago, the remarkable development we described in our chapter titled the *Self* took place.[2] Accepted by most anthropologists to coincide with the earliest evidence of funeral rituals observed in the fossil record, we became a creature conscious of ourselves as if we existed removed from ourselves. We became a being that was self-aware—driven to grasp the nature of our world and to reflect on the meaning of our existence. By about 40 thousand years ago, the last of our archaic forms had vanished, and we set forth in the pursuit of understanding as anatomically modern human beings. Early forms of art, religion, and technology flourished. By about 30 thousand years ago, we had migrated through Indonesia to inhabit Australia and a short time later across the Bering Strait and along the northern ice flows to inhabit North and South America. Over the almost five billion years of the earth's past, humanity had evolved to define the center of life's advance.

---

2. The date of one hundred thousand years as the time of the universe's threshold to reflection is tentative. The advent of the human ability to think and learn reflectively is generally accepted to coincide with the appearance of hunting and funeral rites in the archeological record. Clear evidence of this dates back to ninety thousand years in Neanderthal. Evidence of more anatomically modern archaic forms, however, dates earlier, which suggests that humankind crossed the threshold to reflection earlier. For this reason, the date of humankind's transcendence to reflection will almost certainly change as fossil and archeological evidence increases.

29

Environment

WHEN IT COMES TO the matters of politics, economics, and government, we can't ignore one of the most emotionally charged issues of our time—the *environment*. From a common sense standpoint, what is environmentalism? How over the years has it transformed itself into the cause we know today, and is that cause compatible with a bettering of the human and natural world?

The environmental movement began in the second half of the nineteenth century as a reaction to the near extinction of the bison herds that once roamed the western United States in numbers too large to count. The movement was furthered by the writer Henry David Thoreau and by the naturalist George Perkins Marsh, who, in his 1864 book *Man and Nature*, brought into awareness the idea that we were making disastrous changes to our global environment and that the environment should be allowed to heal on its own or be restored to its natural state. This idea was given additional meaning by the explorer and naturalist John Muir, who felt that natural places should be left undisturbed to fulfill our spiritual needs and those of our children.

By the late 1800s, the views of Muir, Marsh, Thoreau, and others in the early environmental movement led to the first national parks and influenced the outdoorsman and President, Theodore Roosevelt, who in the early 1900s, set aside 125 million acres as national forests. The objective of Roosevelt and of Gifford Pinchot, the first chief of the U.S. Forest Service, however, was practical rather than esoteric. Their goal was to preserve mineral reserves and tracts of timber and grassland for future use and development. The dust bowl and other environmental disasters in the 1930s prompted President Franklin Delano Roosevelt to establish as part of his New Deal the Soil Conservation Service and the Civilian Conservation

Corps. These groups built trails and lodges, reclaimed farmland, and restored damaged ecosystems.

The modern environmental movement began in 1962 with the publication of Rachel Carson's book *Silent Spring*. Carson felt that to create a healthy environment we had to do more than set aside wild areas and create game preserves. She established a link between resources, pollution, and human health. In response, Congress passed the Environmental Educational Act, the National Environmental Protection Act, the Wilderness Act, and the Endangered Species Act—and established the Environmental Protection Agency, or EPA, to monitor the standards set by the Clean Air and Clean Water acts. Western European nations enacted similar legislation; and, in 1975, the United Nations began a program to promote environmental education worldwide.

Many of us have a warm place in our heart for the environmental movement, and even the staunchest critic must concede that environmental activism has benefited our lives. Without environmental activism and the pressures it has brought to bear on business and government, we wouldn't enjoy the quality of air and water that at least in the developed nations we take for granted. But the environmental movement of today is not the movement of Muir, Marsh, Carson, Thoreau, or the Roosevelts.

In the United States, activists have proclaimed goals as far reaching as depopulating the rural west and reintroducing the grizzly bear to habitats in the lower forty-eight states, a move that would effectively close many of the West's most popular wilderness areas to the casual hiker and recreational user. Other groups have burned ski lodges, lumber mills, and sport utility vehicles. Still others have called for eliminating meat and milk from the human diet because cattle and other farm animals release methane, which, arguably, contributes to global warming. On the still more radical end, activists have called for massive population reductions, some even for the planned extinction of the human race. The most extreme views and actions don't represent the environmental movement as a whole. But the movement, as a whole, has become remarkably dogmatic.

In this respect, environmentalism has followed a familiar developmental path, one that tells us much about the present movement. Faced with an issue, we form organizations to pool our resources and political power. As the environmental movement of the 1960s and 1970s shows us, our organizations can allow us to further our agendas. But whether it is a business, government, or environmental organization, when the organization begins to meet the goals it was created to achieve, the focus of the organization invariably shifts. Whereas one might think that an organization would define a new set of goals and channel its energy to their obtainment, it almost always channels its energy to furthering the organization itself. To accomplish this, the organization replaces practical objectives with ideology and uses that ideology to shift its efforts to propagation and to building financial and political support. The organization becomes more important than its goals, which are defined to achieve political rather than practical ends. The Sierra Club wages all-out war on SUVs but pays little attention to high-sulfur coal burning in Asia, a monumental health and environmental concern. Other groups fight the flush toilet and the use of not only disposable diapers but of all diapers. The environmental community has transformed itself from one driven by an altruistic agenda, centered on creating a better place for humanity, to one driven by doctrine and ideology.

Central to the environmental system of beliefs is a conception of nature derived from the idea of the ecosystem as it was popularly defined in the 1960s, a perspective where systems are seen as intricately functioning mechanisms that in their ideal state are in balance. As such, the biosphere is interpreted to be a vast machine where, prior to the ascent of civilization, every species interacted harmoniously with every other species—a complex, in-sync contrivance in an idyllic state of equilibrium.

The biosphere, however, has never been in a sustained state of equilibrium. It has been in a state of evolution. Moreover—since the beginning of the Recent epoch and, with sweeps from the simple to the complex along the way, as far back as the mass extinctions at the end of the Paleozoic era—the net direction of the biosphere's evolution has been toward decline, toward a decrease in complexity and a reduction in the number of species and ecosystems. Paleontologists tell us that in excess of ninety-nine percent of all species that have ever lived have fallen into extinction. Unwilling to incorporate paleontology into their worldview, the environmental movement takes humanity to be that which has destroyed an idealized conception of nature. As such, we are the enemy. We are that which to save the deity of the earth we must confront and in the end defeat.

Today's ideologically driven environmentalist embraces a static, mechanistic view of the biosphere. The objective of the contemporary environmental movement has become to turn back the clock, to return the earth to the ideal state that it was in before "ravished" by the imperfect hand of humankind—to restore the earth to the form "God had intended." Environmentalism has evolved into a dogma, a doctrine, a canon of nature and earth. The movement has transformed itself into a religion embraced with a fervent disregard for common sense.

And for one reason above all others this has been devastating for the environment. Muir, Marsh, Carson, Thoreau, and the Roosevelts saw the environment as inseparable from humanity, as spiritually and biologically nurturing to the human being and necessary for our future. Today's environmentalist sees humanity as irreconcilable with the environment. We are the enemy. We are to be defeated.

Founded by Muir in 1892, the Sierra Club embraced a mission of exploration and preservation. For decades, it built trails and introduced thousands to the natural world. Today, it lobbies the Forest Service to close trails, to stop recreational development, to limit in every way possible our access to wild places. A century removed from the inspiration of its founder, the Sierra Club has recast itself as a legal, political, ideological entity.

For centuries, residents of the American West coexisted with the cougar, or mountain lion. In recent years, political forces have mounted against its hunting, in particular with the use of dogs. The cougar population has exploded and animals too young to remember hounds and hunters no longer fear men and dogs. Cats maul hikers, decimate elk and deer populations, prowl the suburbs for pets, and are driving big horned sheep in California into extinction. Is this best for us? Is this best for the mountain lion and for its environment?

For ten thousand years, Native Americans thinned western forests by setting them ablaze. Settlement ended this practice and led to vegetation buildup, to the massive fires of 1910, and to fire control as a national priority. As Oregon's 500 thousand acre Biscuit fire and other recent burns in the West show us, forests are so overgrown that when fires do start they burn with devastating intensity. In the belief that nature will take care of itself, environmentalists fight grazing, thinning, reforestation, salvage logging, and prescribed burns.

At the 1992 Earth Summit in Rio de Janeiro, the United Nations adopted "Agenda 21," or guidelines for "sustainable development." The core assumption on which Agenda 21 is based is the belief that the human being has no greater place in nature than any other species. Based on this ideal, groups have adopted goals that have ranged from those of clear benefit to us and the environment, such as improving third-world economies, to restrictions on human activity more severe than any George Orwell or the most ruthless communist dictator could have envisioned. In the more radical of these schemes, a global government would oversee a reduction in the earth's population from today's six billion to a few hundred million. Human habitation would be limited to "green" communities, and the majority of the planet would be off-limits to human entry. The government would eliminate private property, ration water and energy, and control housing, industry, agriculture, and transportation. Pregnancy would require a permit. Education would be governmentally sanctioned and environmentally based.

Not for millennia has the natural world existed without the human world, and we're not going away. We can improve forestry, manufacturing, and other industrial practices. We can build cars, roads, and homes that better interact with our environment, but we can't eliminate ourselves from the environmental equation. Capitalism and scarcity-based economics drive environmental exploitation, and we need an environmental movement to counter. For the good of humanity and for the good of the earth, environmentalism must put common sense before dogma and inspire a new generation of Muirs, Marshs, Carsons, Thoreaus, and Roosevelts.

30

Population

IN THE ENVIRONMENTAL MOVEMENT, a long-standing area of concern has been the earth's population and its growth, estimated by the *United Nations Population Division*[1] to be currently about six billion and, by medium projections, to increase to about nine billion by the middle of the century. As with most environmental issues, the matter of population isn't as simple as the environmentalists would like us to believe: "less people good, more people bad." It's not morally or politically possible to reduce the earth's population to the between three hundred million and two billion groups proclaim as the most the planet can sustain, which would do away with just about everyone but the environmentalists. To make sense of the many "facts" about population and population growth we come across every day, a look at population demographics is necessary.

According to the March 2004 bulletin of the *Population Reference Bureau*, an organization that has been tracking population demographics since 1929, the earth at the start of the twentieth century had a population of about 1.6 billion people. By the end of the century, that number had reached 6.1 billion, with most of this growth taking place after 1960. In line with United Nations projections, the bureau estimates a global population of about 7 billion by 2015.[2]

As one would expect, two basic factors influence population dynamics: birth and death. Interestingly, scientists don't attribute the rapid growth in the world's population during the latter half of the twentieth century to an increase in the birthrate. They attribute it to a lowering of the death rate brought on by an increase in life expectancy. In 1950, an individual in the undeveloped world lived about 41 years and in the devel-

1. "World Population Prospects The 2002 Revision," 1.
2. "Transitions in World Population," 3.

oped world about 66 years. By the year 2000, these figures had increased to respectively 63 and 76 years.[3]

This stretching-out of the lifespan gave rise to what those in population circles call a *demographic transition*. Throughout history, human population has grown rather slowly. Women had a lot of children, but high birthrates were offset by high death rates from war, disease, poor nutrition, and poor sanitation. Modern medicine and agriculture and above all extensive sewage and water treatment lengthened the average lifespan. Birthrates exceeded death rates to such an extent that the earth's population exploded. But this doesn't mean that the earth's population will always grow. Life expectancy must at some point level off; and, as we see in the United States, where by some measures the number has slightly declined, may have reached a plateau at between 76 and 78 years. As life expectancy peaks and the death rate catches up to offset the birthrate, population growth will level off.

This takes us to the opposite side of the population equation. Though the primary variable in today's demographic transition is the death rate, the birthrate also comes into play. In contrast with the death rate, however, the worldwide birthrate, or fertility, the last couple of centuries has plunged.

In the early 1800s, the average American woman gave birth to seven children. By the early 1900s, this number had decreased to four children. Today, the average American woman gives birth to about two children, and the birthrate in Italy, Spain, the Czech Republic and other European nations has dropped to as low as 1.3 children. Even in the developing world, women are opting to have smaller families. In the 1950s, the average woman in Kenya gave birth to 6.7 children. Today she gives birth to 5.2 children. In Asia over the same period, the number has decreased from 5.9 to 2.6 children, in Latin America and the Caribbean from 5.9 to 2.7 children.[4]

Declining fertility has many causes. In the developed world, it's the result of delayed marriage, higher divorce rates, and woman choosing to attend college and to opt for career over or in addition to family. As significant, living standards for many working families in the United States and other developed nations have declined over the last decades. Couples

---

3. Ibid, 5.
4. Ibid, 7–10.

that want large families can't afford them or have to spend so much time on the job and away from the home that they can't properly raise them and, responsibly, have opted for fewer children.

In the undeveloped world, a parallel set of dynamics has come into play. When countries modernize and women become better educated and adopt new roles and lifestyles, families get smaller. In Mali, women with no education have an average of 7.1 children. Women with secondary or higher education have an average of 4.1 children. In Peru, women with no education have an average of 5.1 children. Women with secondary education have an average of 2.4 children and women with higher education, 1.8 children[5].

In no small way, the desire of women throughout the world to have smaller families has been made possible by contraceptives. The pill, the intrauterine device, and other methods have allowed couples to plan for rather than accommodate pregnancy. According to the United Nations Population Division, more than 71 percent of married women in Europe and the United States use contraception during their reproductive years. The number of fertile women in Kenya on some form of birth control has increased from 4 percent in the 1970s to about 32 percent today. In Bangladesh, the number has increased from 5 to 43 percent and in Columbia from 9 to 64 percent. Worldwide, almost 60 percent of married women engage in family planning.[6]

Periodically, the United Nations Population Division publishes population projections through the year 2050. In a high growth scenario, which is essentially a linear projection of the current growth rate, the earth will be inhabited by 10.6 billion people in 2050. In a more realistic medium growth scenario, it will be inhabited by 8.9 billion. In a low growth scenario, which better takes into account declining fertility, population growth will peak at about 7.6 billion in 2037 and decline to about 7.4 billion in 2050.[7] Over the years, the Population Division has revised their estimates downward as fertility and birthrates have become better understood. This suggests that the earth's population could reach its maximum sooner than the 2037 estimate.

---

5. Ibid, 17.
6. "Levels and Trends of Contraceptive Use as Assessed in 2002," 8.
7. "World Population Prospects The 2002 Revision," 1.

When, in the not-too-distant future, lifespan in the developed world peaks and lifespan in the undeveloped world catches up, the earth's population will begin what some predict will be a sharp decline to at some point settle into a neutral growth pattern where birth and death rates are equal. But this doesn't mitigate the concerns of the population demographics we face today and will face in the years immediately ahead.

In the developing world, population will continue to grow for the next few decades. This will place further demands on farmland, natural resources, and living standards. This pressure is compounded by another factor—AIDS. In a country where the population is increasing, we find a pyramidal distribution of the population between the ages: many young people, less in middle age, and still less in old age. In nations with a high birthrate like Kenya, AIDS has distorted this population distribution in a devastating way. The disease has killed many people in middle age, those who are the most productive members of society and on whom the old and young depend, and left the country with a vast number of orphaned children. Teenagers are often compelled to care for themselves, their grandparents, and their younger brothers and sisters.

Though nowhere near as overwhelming as in Africa and other undeveloped regions, the world's industrial nations also have population concerns. China, Russia, the United States, and almost all European nations have a birthrate below the 2.1 children per woman needed to maintain their populations. In a nation with a falling birthrate, there will be a population distribution with a small number of children and a large number of elderly with respect to those in middle age. This distribution threatens to make insolvent the many elderly and other entitlement programs the world's developed nations maintain. When founded by Franklin Delano Roosevelt in 1935, Social Security in the United States had in excess of 50 workers paying into the system for every recipient. By the 1950s, this figure had dropped to 16 workers paying into the system for every recipient. Today, it runs at about 3.3 workers per recipient, and by the end of the decade it will by some estimates drop to about two workers for every recipient.

Nations have had almost no success encouraging women to have more children and to forestall population decline have opened their borders to immigration. This has created a situation where the native generations that built the country are declining in number with respect to an immigrant population. In the United States, tension over the issue has grown in recent years, but the country has thus far been able to cope. As a nation of immigrants, the United States is culturally diverse, and migrants come largely from Mexico and South America and have similar work and religious values. Immigrants to Spain, France, Germany, Great Britain, the Netherlands and elsewhere in Europe come predominantly from Islamic North Africa. The descendents of civilization's founders and as potentially powerful a force for human good as any, they're arguably less accepting of cultural differences than native populations. Values and expectations have and will continue to clash, and radical members of the European Islamic community have gone so far as to call for the reforming of European nations into Islamic states.

Ironically, the issue of overpopulation so often the focus of environmental activism, which must be credited with bringing the topic to the forefront of thought, will take care of itself. In the future, we won't be concerned about population expansion but about our transition from a post 1960 era of rapid growth—and the social, political, and governmental structures created in response—to a twenty-first century era of declining population and an eventual settling into a neutral growth pattern, at whatever level we deem best or that might turn out to be.

31

Global Warming

JUST AS THE ENVIRONMENTAL movement is reluctant to move beyond the politically successful idea of overpopulation—and when it comes to family planning and other goals with tangible benefits should continue its efforts—it clings to many scientifically challenged views. So extreme is environmental doctrine that in a 2002 study, the *World Wildlife Fund* reported that if the human race continued to "plunder" the earth at its present rate, our planet would be dead by 2050 and we'd be forced to colonize two additional planets.[1] In our supposed plundering of the earth, no environmental issue is more controversial than *global warming*.

To understand how our climate may today be changing we need to look at how our climate has in the past changed. To assess past temperature and precipitation, climatologists use a variety of geological principles and technologies. Important are core samples taken from the ocean floors. When microscopic plants and animals in the water die, their shell remains settle on the bottom. When studied in a core sample, these remains reveal much about past temperatures and other conditions. When core samples from around the world are correlated and combined with core samples taken from the polar icecaps and with other geological evidence, they help to establish a climate record that dates back almost to the earth's formation.

This record tells us that the earth's average surface temperature has remained remarkably constant, 20° C or 70° F, since the advent of life about four billion years ago. Although on the average, temperature fluctuations cancel out, there have been ups-and-downs. In general, the earth has experienced long periods of warm weather punctuated by short periods of cold weather. These cold periods, or *glacial episodes*, occur every 150 or so million years and last a few million years. The longest of the

---

1. Townsend, "The Earth Will Expire by 2050."

earth's near ice ages, the *Permo-Carboniferous*, began about 300 million years ago. The earth's most recent glacial period, the *Quaternary*, began about 2.5 million years ago. Although continental ice sheets withdrew between ten and thirteen thousand years ago, most geologists think the earth is still in the Quaternary ice age.

The reason for this is that our climate not only experiences cycles of long warm periods broken by ice ages but fluctuations within ice ages. Each ice age has glacial advances and retreats. Within these glacial cycles are yet other climatic fluctuations, and within these still others. Prior to the Egyptian Empire, the earth was warmer and wetter than today, the Sahara Desert much like the savannah of East Africa. During the Early Middle Ages, temperatures were cooler than today, a factor believed to have contributed to the turmoil of that period. During the High Middle Ages temperatures were warmer. In Great Britain, grapes grew several hundred miles further north than they do today, a situation that heightened tension with the French. In the 1500s, the world plunged into the *Little Ice Age*.[2] Glaciers overran towns in the Alps. Farmers shifted from grain to potato based agriculture. Vikings abandoned farms and settlements on once arable shores in Greenland. In his march against Russia, Napoleon confronted one of the most brutal winters on record. In the mid-1800s, temperatures began to warm, a trend that continued to about 1940. Between that date and 1970 temperatures remained constant or slightly declined. In 1980 they began to increase, to peak in 1998 and remain constant or somewhat decline to the present. Even in our lifetime, we experience cycles of rain and drought and of warm and cold temperatures. In a glacial period, there's no such thing as a "normal" climate.

This brings us to the issue of cause. The variables behind climate change aren't well understood, but scientists agree on one thing. No single factor accounts for climatic fluctuation.

Astronomical variables affect how much energy reaches the earth and how much the earth reflects into space. Our galaxy rotates once every 300 million years. As with tidal processes, two phases appear to exist for this cycle and to coincide with 150 million year glacial events. In addition, the sun has cycles of low and high energy output, and the earth's orbit

---

2. What is popularly referred to as the Little Ice Age defines a period when there was a great deal of climatic fluctuation. For this reason, the dates at which the Little Ice Age began and ended are open to interpretation. The period between 1500 and 1850 is widely accepted.

around the sun is elliptical, which varies the energy received from the sun on a 93 thousand year cycle. Moreover, the equator tilts with respect to the orbital plane on a 41 thousand year cycle, and the earth has a precession cycle, or wobble, that lasts 26 thousand years.

Also important is the influence of geological activity on climate change. Continental drift and the uplift of mountains alter winds and currents. The earth experiences cycles of volcanic activity and fault line movement that correspond with cycles of increased and decreased energy transfer from the interior to the surface. Volcanic eruptions alter levels of atmospheric gases and particulates; which, as the eruption of Krakatau in 1883 and the documented cold winters that followed illustrate, can reflect solar energy into space. Scientists have postulated that ocean salinity is a variable, as is the strength and position of the earth's magnetic field.

By far the most complex and least understood variables of climate change are those associated with the biosphere. The average temperature of our planet has remained remarkably constant since the advent of life, and the biosphere has played a clear role in maintaining this consistency. Forests grow and shrink. Deserts expand and contract. Algae concentrations in the world's oceans increase and decrease. In ways we've scarcely begun to understand, the biosphere embodies the feedback mechanisms to sustain life.

Only when we take into account fluctuations in energy sources, fluctuations in the planet's ability to harness energy, and the biosphere's ability to buffer these fluctuations, do we begin to appreciate the complexity of our planet's climate.

In contrast, there's the theory of "global warming." For a moment, let's set aside our political allegiances and look at the issue from the standpoint of common sense.

Energy constantly reaches the earth from the sun, and the earth constantly radiates a portion of that energy into space. As long as a balance is struck between energy inflows and outflows, the earth's temperature remains constant. In the laboratory, however, certain gases have been shown to act like the glass on a greenhouse. They let more energy in than they let out, at least with respect to certain wavelengths of light.

With the exception of water vapor, which is by far the most potent and common greenhouse gas—remember two-thirds of the world is covered by rivers, lakes, and oceans—there are two primary contributors: carbon dioxide, released in metabolism and from the burning of fossil fuels, and methane, synthesized in any number of biological processes but often cited by environmental and animal rights groups as produced in the digestive tracts of cattle. According to global warming theory, industrial, agricultural, and other practices release greenhouse gases into the atmosphere, which trap heat and increase the earth's temperature.

And there is evidence to support this theory. Measurements suggest that atmospheric levels of methane and carbon dioxide and, at least when taken as an average, the earth's temperature have increased over the last one hundred years. But widespread global metrological monitoring only began during World War II, and even the best measurements are open to interpretation and like any data can be manipulated to support an argument. Not often pointed out by "global warming" advocates, for example, the earth's temperature rose at the end of the Little Ice Age, a trend that began more than one hundred years before widespread industrialization.

The most often cited evidence for global warming comes from computer models. In these models, scientists attempt to quantify the variables that affect the climate and to predict what happens when one or another change. But what was the sun's energy output at a given time in the past? How do biological processes buffer humidity and temperature? Do methane and carbon dioxide have the same greenhouse effect in the atmosphere as in the lab? Because the variables associated with climate change are so complex, researchers are forced to make assumptions. Particularly speculative is the sensitivity of the climate to changes in carbon dioxide levels. With one set of assumptions, a climatologically model tells us to prepare for global warming, with another for an ice age—a view widely held by the environmental community in the 1970s and embraced by some today. In 2006, Democratic Party Chairman Howard Dean stated the belief that human activities cause global warming, and global warming will cause the earth to enter into a new ice age. Even a figure as basic as the earth's average temperature is remarkably difficult to determine, and many scientists feel that until the 1970s and the widespread use of weather satellites we had no reliable way to compare year-to-year global temperatures. The assumptions of cloud behavior incorporated into computer models are also controversial and can have a significant impact on results.

Also controversial in global warming theory is the effect of a temperature increase on our lives and on the environment. Global warming's leading advocate, former Vice President Al Gore, recently told us that if we don't drastically cut greenhouse gas emissions in the next ten years it will be too late and the consequences for humanity and the earth will be dire. The issue as to how anyone could come up with a figure like Gore's ten-year window aside—and the science, the lack thereof, and the agendas of politicians and advisors this implies—is a warmer climate necessarily a bad thing? The High Middle Ages was a warmer period than today, and it marked one of history's most dramatic stages of advance. Barring the massive increases in sea level and other catastrophic changes adored by Hollywood and that in reality are highly speculative—Venice, which was founded in the fifth century, was not under water during the High Middle Ages—certain parts of the world would face changes we would consider detrimental and others would face changes we would considered beneficial.

Climate change is accounted for by a multitude of geological, biological, and astronomical variables, many of which aren't clearly defined or well understood. Global warming theory disregards this complexity in favor of suppositions that lead to a simple, cause and effect model: human activity produces global warming.

The environmental movement embraces a static view of nature: the premise that in its ideal state the earth's climate is constant. In light of the Quaternary's 2.5 million years of glacial advances and retreats how can the earth's climate not be changing? Our love of SUVs and prime rib may be a factor, but the effect of our activities would be imposed on the forces that have long driven climate fluctuation. Thirteen thousand years ago, most of the eastern United States—Boston, Chicago, New York City—was under a continental ice sheet that in places was more than a mile thick. Think back to our chapter titled *Science*. To know what if any impact human activities have on the climate, we would have to design an experiment that isolated natural from artificial variables and that accurately quantified these variable. Given the complexity of the earth's climatic system, this may not be possible. Human caused, or anthropogenic, global warming is not and may not ever be a scientifically verifiable hypothesis. Aware of this, the environmental community has politicized the issue to such an extent that global warming—or "climate change" as, in light of more than a decade of stable or cooling temperatures, environmentalists now call it—is taken to be a fact. This places the burden of proof against it on its opponents and grossly distorts the methodology of science—forces the researcher to disprove an unproven hypothesis. As unsubstantiated and politically motivated as the theory of global warming may be, however, it compels us to face a situation worthy of our attention. Cause aside and for warmer or colder, our climate has changed, is changing, and will continue to change.

32

Cap and Trade

IF IT WERE PURELY a matter of science, the issue of global warming would be a curious diversion in our quest to understand the workings of the biosphere. Climate models, the variables associated with climate change, and the way they interact and feedback on one another are truly fascinating, and their understanding represents a remarkable intellectual effort. As it stands, the issue of global warming, or human caused climate change, embodies less science than politics. This invokes a cost, one that will be paid by every American. The primary tool through which our elected officials propose to strap us with the bill is called "cap and trade."

To understand cap and trade and its impact, we begin with a look at an international treaty called Kyoto—or to be accurate the Kyoto Protocol to the United Nations Framework Convention on Climate Change in December of 1997, named after the Japanese city in which it was proposed. Established on the premise that greenhouse gases emitted by human activity are causing a significant and disastrous change in the earth's climate, the Kyoto Protocol calls for nonexempt signatory nations, primarily the United States and the countries of the European Union, to, by between 2008 and 2012, reduce their emission of methane, carbon dioxide, and four other greenhouse gases to a level that is 5.2 percent below a 1990 baseline. The principle method called for in the treaty to accomplish this objective is to establish a market to buy and sell emission rights.

At its core, environmental legislation regulates human activity and behavior. Environmental activism is centered on the belief that an "enlightened elite" must impose what it deems to be for the good of the earth on a population that would not otherwise modify its behavior. As over the previous century the environmental movement developed, it adopted two principle methods to accomplish this objective.

First, by way of government, we impose mandates, or, through the legal system and the police power of the state, directly regulate human activity and behavior. We set standards for air and water quality and for factory and vehicle emissions. We establish allotments for timber harvest and limit areas open for oil and mineral extraction. By way of government, we dictate and enforce environmentally correct conduct.

Second, we manipulate the market forces of capitalism. This approach is predicated on a view of environmentalism widely held by economists in the 1960s and 1970s. It's the belief that environmental problems are the result of an improperly functioning economy, one in which supply and demand fail to assign the "correct" price to commodities. Gasoline, for example, should not be priced based on the cost of refining, crude oil, and the amount people are willing to pay but should include "intangible" costs: the cost of wildlife habitat lost to road construction, the cost of cancer and emphysema caused by air pollution. If government added these costs to the price of gasoline by imposing a tax, gasoline would be more expensive. Demand would drop. We would drive less, and the automobile would have less environmental impact. By adjusting the price of commodities, or incentivizing certain activities and placing obstacles to others, human behavior would align with the needs of the environment, as those in power define these needs. By government manipulation of the economy, man and earth would exist in balance.

Inspired by the Kyoto Protocol, cap and trade incorporates both approaches to environmental legislation. In the cap and trade scheme, a company or other entity is allowed to emit a governmentally determined amount of greenhouse gases—chiefly carbon dioxide. If a company emits more than allowed, it must buy "carbon credits" to offset its emissions from an emissions trading market. If a company emits less, it's free to sell its unused carbon credits on the market. Over time, standards would be lowered, in theory reducing greenhouse gas emissions. In effect, cap and trade creates an artificial scarcity of fossil fuel energy. Complexity aside, it does what a gasoline, carbon, or any other type of energy tax would do, increase price based on the supposition that it will reduce demand.

As proposed, not only would industry be required to purchase carbon credits, individuals would be required to purchase carbon credits. As they boast that they now do, Al Gore would have to buy carbon credits to offset the fuel his Gulf Stream jet burns while flying him around the world to promote energy conservation and the reduction of greenhouse gas emissions. California Governor Arnold Schwarzenegger would have to buy carbon credits to offset the fuel his private jet burns on his daily commute from his home in Los Angeles to the state capitol in Sacramento. You and I would also have to buy carbon credits. If, say, you have a large family and needed an SUV, or live in a rural area or worked in construction and needed a pickup, and that vehicle failed to comply with a fuel efficiency standard, you would at the time of purchase, or with each stop at the pump, have to buy a carbon credit. Great Britain has implemented a similar plan. On the purchase of a new vehicle, not only must the consumer absorb the usual fees and taxes, he or she must absorb the cost of what can amount to a large carbon offset. Carbon credits would also apply to older, less energy efficient homes and buildings, and in theory can be incorporated into almost every transaction, as almost every good and service involves some degree of fossil fuel consumption. New Zealand has even proposed that a land owner buy a carbon offset to clear brush or timber from his or her property.

As one may surmise, and as the British experience has shown, cap and trade schemes are difficult to implement and have a plethora of ramifications. Environmentalists point out that by making "pollution" a commodity that like any other product can be bought or sold, it destigmatizes the act of pollution and makes it politically acceptable. Companies may also work within the tangle of laws and regulations cap and trade has spawned, and is certain to further spawn, and avoid compliance. Industries may lobby for exemptions. Companies may buy or borrow credits to be allotted in upcoming years—establish a futures market in carbon credits. An American company may buy credits from a less environmentally correct foreign company or one set up to circumvent the cap and trade laws. If say a Chinese company decides to sell its carbon credits and not reduce its emissions or is purely a paper carbon-trading entity, what really can we do about it? In cap and trade plans like the ill-fated America's Climate Security Act of 2007—Senate Bill 2191 sponsored by Senator Joe Lieberman, an Independent from Connecticut, and Senator John Warner, a Republican from Virginia—to ensure the cooperation of foreign companies we would embargo non-complying products. The bill, however, makes no provision to determine which products are made by foreign companies not meeting carbon standards. The government can't protect the public from poisonous pet food and lead-laced children's toys imported from China. How many "carbon emission" inspectors would it have to hire to even begin to implement an enforceable cap and trade scheme?

When we take into account the impracticality of enforcement, will cap and trade have any impact at all on carbon dioxide levels? Cap and trade is a legal, political contrivance. Carbon dioxide emission credits are an artificial construct. Al Gore and Arnold Schwarzenegger's private jets still burn fuel, even though their purchase of carbon credits may appease political factions and make them feel better about it—not to mention employ a lot of accountants to shuffle the carbon credit paperwork and jiggle the numbers between accounts. China has overtaken the United States as the world's largest emitter of carbon dioxide, and the growth of emissions in China and in India is expected to increase.[1] Dr. Patrick Michaels, former president of the American Association of State Climatologists and now senior fellow in environmental studies at the Cato Institute had this to say with regard to the Lieberman-Warner Act: "Say the United States actually does what the law says, though no one knows how to. The result is an additional 0.013 degrees centigrade of 'prevented' warming."[2] He then goes on to explain that statistically this is a meaningless figure in that it is a fraction of expected annual variations.

Yet again, there is another side to this argument. To actually offset carbon emissions, for every gallon of fuel Al Gore's jet burns, someone, you and I, would have to burn a gallon less fuel. Proponents point out that we can plant trees and use wind and solar power to offset emissions. Realistically, however, how many trees can we plant; and, as we'll look at in an upcoming chapter, wind and solar power have practical limitations. Cap and trade may reduce carbon emissions, but the only significant way it will do so is to increase the cost of energy. Depending on the specific cap and trade plan, and depending on the source of the statistics and the assumptions one makes concerning the implementation of new technologies and non-carbon energy sources, cap and trade would increase the cost of gasoline, electricity, natural gas, and heating oil by a minimum of between 55 and 145 percent—adding a at least two additional dollars onto each gallon of gasoline. According to the National Association of Manufacturers, which conducted a state-by-state study of the impact of the Lieberman-Warner Act, the cost of gasoline would rise by as much

1. Lieberman, "Five Myths About the Lieberman-Warner Global Warming Legislation."
2. Michaels, "Cato Scholar Comments on Warner-Lieberman Climate Security Act."

as five dollars per gallon.[3] The most recent cap and trade plan, a massive 1200-page bill proposed by House Majority Leader Nancy Pelosi and supported by the Obama Administration, is even more sweeping. In testimony before the House subcommittee on Income Security and Family Support, Congressional Budget Office senior advisor Terry Dinan said that a 15 percent cut in carbon dioxide admissions, a fraction of that proposed in the Obama plan, would cost the average American household 1,600 dollars per year and as much as 2,200 dollars.[4]

As we observed in 2008, an increase in fuel price can reduce demand for energy. Faced with four-dollar-per gallon gasoline, Americans drove their SUVs and trucks less, and demand for compact and subcompact cars increased. But how far can Americans cut back on energy consumption? We aren't Europe, and not even Europe has been able to meet the cutbacks required in the Kyoto protocol. America is expansive, and the design of our cities is different. Our geography doesn't lend itself to extensive public transportation systems; and, with the exception of some urban areas, we can't give up driving altogether. Americans have to drive to work, to town, to school. The goal of cap and trade is to reduce allowable emission amounts over time. In a recent speech, Representative Greg Walden, a Republican from Oregon, reported that the Obama cap and trade proposal called for a 20 percent reduction in greenhouse gases from their 2005 levels by 2020 and an 83 percent reduction by 2050. This would cut the United States consumption of fossil fuels to the level of 1910 or to that of present-day Nigeria.[5] Without replacement by a non-carbon based energy source, and not taking into consideration an expected increase in population, this translates into an 83 percent reduction in economic activity—which translates into an 83 percent reduction in living standards. How much will the economy stand? How much are we willing to give up?

The average American is the loser in the cap and trade scheme. Who benefits?

Above all, it's the companies, governments, and other organizations that are or will operate the carbon markets and profit from the exchange of carbon credits. Until recently, the leading lobbyist for cap and trade

---

3. "Analysis of The Lieberman-Warner Climate Security Act (S. 2191) Using The National Energy Modeling System (NEMS/ACCF/NAM)."

4. Dinan, "The Distributional Consequences of a Cap-and-Trade Program for CO_2 Emissions."

5. Varble, "Walden calls Obama's energy bill 'an Oregon job killer.'"

was the now defunct energy trading group Enron. Al Gore buys his carbon credits from a London based firm named Generation Investment Management, a company that he in part owns and for which he serves as Chairman.[6] An outcome of the Kyoto protocol, the United Nations has positioned itself to operate a world-wide carbon exchange, and—as with its failed attempt to take control of the Internet, tax users, and regulate content—sees it as a vast source of funds and power. Proponents point out that cap and trade proceeds will be reinvested in non-carbon based energy sources. By the most generous assessments, only a fraction of revenues would be directed toward this end. The Obama administration targeted cap and trade as a source of funds to pay for its proposed "middle class" tax cuts. Legislators see it as the largest tax increase in United States history—a vast source of revenue needed to prop up the nation's failing Medicare, Social Security, and other entitlement programs.

The second major benefactor of cap and trade is the ideological wing of the environmental movement—those who envision an "Agenda 21" or other radical "green" future. As we'll discuss in our chapter titled *Renewable Energy*, wind and solar power—the only renewable sources of energy given serious consideration by the Obama administration—have significant engineering limitations. The only way "green" energy can become cost competitive with conventional sources of energy is to drastically increase the price of fossil fuels.

The validity of the global warming premise aside, cap and trade illustrates a point we should all be aware of—the unintended consequences of our actions. There are politicians who truly care about the environment. Al Gore may sincerely believe in global warming and its dire consequences. No matter the agenda we choose to further, however, we must step back. We must evaluate our cause in light of the consequences it may invoke. By any commonsense assessment, the only impact cap and trade would have on the environment would be due to a massive reduction in economic activity, an artificially invoked recession that by some estimates would dwarf what we now face or any in history. So blatant is the downside to cap and trade that proponents no longer call it "cap and trade." Just as the label of "climate change" is an attempt to put a good face on global warming, the labels of "Cap and Cash Back" and "Cap and Pollution Reduction" attempt to put a good face on cap and trade. No matter our political allegiances,

---

6. *"The Heat is On: Gore's 'Carbon Offsets' paid to a firm he owns."*

we not only have a responsibility to further our agenda, and whatever financial or other benefits it may bring us, but to weigh our cause against the totality of its impact—on our nation, on our neighbors, on our way of life. As we'll discuss in upcoming chapters, even if one accepts the hypothesis of anthropogenic global warming, there are commonsense ways to limit the emissions of greenhouse gases, solutions that allow for economic growth and individual freedom rather than place yet another level of government control on our lives.

33

Energy

INSEPARABLE FROM THE ISSUE of global warming and cap and trade is the issue of *energy*. In the minds of some, energy is a contributing factor to nearly all of humankind's social and environmental ills. As politicians argue for or against this or that energy policy, one point stands out. Most don't know the first thing about energy.

The energy we use comes from two sources: the sun and the earth.

Solar energy drives winds and ocean currents and lifts water from the oceans and drops it onto the land where it runs back into the oceans. Plants use sunlight to synthesize carbon and hydrogen into sugars, amino acids, and carbohydrates, and animals harness the sun's energy by consuming plants.

And wherever we can do so practically, we tap into the flow of solar energy that strikes the earth and that works through its systems. Solar collectors and photoelectric cells convert sunlight into heat and electricity. Turbines and generators convert the movement of air and water into electricity. We burn garbage and wood waste to produce steam and turn generators that produce electricity.

But our most important source of solar energy doesn't come from sunlight that strikes the earth today; it comes from sunlight that struck the earth long ago. In the past, particularly during the Carboniferous Period of the Paleozoic Era, life was more plentiful than today, and over the course of hundreds of millions of years locked up vast amounts of solar energy in its organic structure. In the form of oil, coal, and natural gas, these *hydrocarbon* remains, or *fossil fuels*, are our most important source of power.

But we don't rely entirely on these or other forms of direct or indirect solar energy. We also harness energy from the earth. The earth is hot inside, its temperature in part the result of the decay of uranium and other radioactive elements. In places where the earth is warm near its surface, we harness this heat directly, *geothermal energy*. More commonly, we draw energy from the earth by harnessing the reactions that heat it—*nuclear energy*.

We've long understood solar and earthly sources of power, and each year we increasingly tap these sources. Today, the average citizen in the developed world consumes more energy in a year than a member of a preindustrial society did in a lifetime.

And, as from the industrial revolution to the present we have increased our reliance on technology and our consumption of energy, we have realized that every use and source of power has its problems and ramifications.

Electricity is generated in power plants and distributed through a *grid*. Some plants are driven by wind, water, and sunlight. But there are only so many rivers to dam, and at the sunniest and breeziest locations solar and wind generators produce electricity only about eight hours a day. Most plants are fueled by oil, coal, or natural gas, which bring real pollution concerns, such as sulfur emissions that cause acid rain and mercury emissions that contaminate fish, and perceived or at least controversial pollution concerns such as global warming.

Fossil fuels also create geopolitical concerns. The world's largest petroleum reserves are in unstable parts of the world—in particular North Africa and the Middle East. So heavily do industrial nations rely on imported oil that an extended interruption in flow could thrust the United States and other petroleum consuming countries into depression, even anarchy. Rhetoric aside, military and diplomatic efforts in the Middle East and other troubled parts of the world have one overriding objective—to maintain the flow of oil.

The most abundant source of energy we know how to harness is nuclear power, or nuclear *fission*. Nuclear fission, however, brings the concern of radioactivity. Proponents point out that trace radioactive elements are all around us and that each year we release more into the atmosphere by burning coal than we would release in the worst nuclear accident or waste spill.[1] Opponents point out that in a nuclear accident or waste spill radioactivity is concentrated and, as in the Chernobyl disaster of 1986, could make areas uninhabitable for years.

---

1. Gabbard, "Coal Combustion: Nuclear Resource or Danger."

Another form of nuclear power, *fusion*, produces almost no radioactive byproducts. In fusion, two hydrogen atoms combine to form one helium atom, releasing a tremendous amount of energy. This reaction powers the sun and hydrogen bomb. But a fusion reaction is so difficult to initiate and contain that after a half-century of research scientists haven't been able to sustain a controlled fusion reaction for any practical length of time.

Baring the perfection of fusion or the development of some unforeseen energy technology—and such may not be totally unrealistic, who two hundred years ago envisioned nuclear power?—every use and source of energy has its complications.

But we need energy, and we're always looking for ways to better use and develop our energy resources and to minimize their economic, ecological, and political impact.

With mixed success.

Government periodically sets higher automobile gas mileage standards. We improve mileage in three principle ways: by reducing the car's weight, or the amount of mass we accelerate and decelerate; by reducing the car's drag, or its wind and other resistance to movement; and by increasing the efficiency of its engine. There's not a lot we can do to decrease a car's drag. Not every car can be as sleek and roll as easily as a Corvette or Porsche roadster. In its present internal combustion form, engine efficiency may be near its limit and smaller cars aren't always safe and practical. When gas mileage standards in the 1970s forced the full-sized station wagon off the market, families in the United States turned to the SUV. Classified as trucks and thus exempt from strict mileage requirements, they offered families the large vehicles they needed, albeit vehicles that for the most part were less safe and efficient than the old station wagon. Environmentalists have long touted the electric car as the solution to our energy and pollution problems. But electricity has to come from somewhere. By the time it's generated, transported, and stored, an electric car can burn as much or even more fossil fuel than a gasoline car.

Duel fuel vehicles, or cars that manufacturers advertise as able to burn ethanol or gasoline, are another political rather than practical solution. Alcohol has a lower energy content than gasoline. To in part compensate, an alcohol fueled engine must have a higher compression ratio, a characteristic of an engine's bore, stroke, and combustion chamber size that even with turbochargers and variable valve timing can't be readily changed after assembly. To be able to burn gasoline, therefore, a dual-fuel engine must run with a low compression ratio. Consequently, when burning ethanol, mileage will be lower—a drop of up to thirty percent depending on the gas-ethanol mixture. As methanol powered race cars have over the decades proven, to run efficiently an engine must be constructed to work with a specific fuel. The exception is bio-diesel, which, when properly formulated, is chemically similar to mineral diesel and thus can be blended with or used in place of the conventional fuel.

Producers of "renewable" energy such as ethanol made from corn and bio-diesel made from soybeans claim they have the solution to our energy problem. They fail to mention that although another form of alcohol, methanol, and cellulosic ethanol, or ethanol made from plant fibers, may have a role in the future; bio-diesel and in particular corn-based ethanol take almost as much energy to produce as they yield. Al Gore and the global warming movement call for a massive government program to replace fossil fuels with wind and solar power. They, however, fail to take into account the physics of wind and solar power generation. As we'll discuss in our chapter titled *Renewable Energy*, from an engineering standpoint, wind and solar power have practical limitations.

It's trendy to think of hydrogen as the fuel of the future. What advocates neglect to tell us is that there is no source of chemically unbound hydrogen on the planet. There's a lot of hydrogen, but it's tied up in water and hydrocarbons, which means that to release it we have to input energy. And it takes more energy to break apart a water or hydrocarbon molecule than we get back when we burn the hydrogen. Unless we turn to nuclear power, the hydrogen fuel cell cars said to be the wave of the future will do little more than shift the use of hydrocarbons from the automobile to oil, gas, and coal powered hydrogen extraction plants and increase our reliance on fossil fuels.

Though it may seem like every source of energy has significant limitations, there are practical energy production and conservation technologies. One is hybrid gas and electric cars, which, though they may burn as much fuel as conventional cars in certain conditions, burn less in stop and go driving. Because many of us may be considering the purchase of a hybrid vehicle, we'll take a closer look at the technology in an upcoming chapter. The most tried-and-true energy conservation technology is diesel. Due to the energy content of diesel, high compression ratios, direct injection, and the use of turbochargers, diesel engines operate at about thirty percent greater efficiency than gasoline engines. Diesel hybrid combinations are particularly attractive. European nations are adopting diesel, where the fuel powers about seventy percent of new cars sold. In the United States, the environmental movement has long fought diesel's use, and California has proposed what amounts to an outright ban on certain types of diesel engines. Though diesel produces less pollution overall, it, arguably, produces higher particulate emissions. Recent emission requirements have also substantially decreased diesel efficiency and substantially increased fuel costs and carbon dioxide emissions. On the engineering side of diesel's limitations, only so much diesel can be distilled from a barrel of oil. Due to demand in China, India, and Europe, there are periodic, world-wide shortages of diesel and similar density jet and other fuels. This suggests that at least in the immediate future, gasoline powered cars and trucks will continue to dominate the United States market.

Energy comes from the sun and the earth. Our demand for energy has grown and by most projections will continue to increase. In this chapter, we provided an overview on energy uses and sources and learned that every form of energy has its benefits and its drawbacks. In the chapters that follow, we'll look at the technologies that in the future will play the greatest role in our lives. There are commonsense solutions to the world's energy problem. In certain respects, however, they may not be what we might expect.

34

Hybrid Cars

RECENTLY, SOARING ENERGY COSTS made the hybrid car an automotive rage. Dealers couldn't keep them on the lot, and people were willing to pay thousands more for the technology. Because many of us may be considering the purchase of a hybrid car, a brief look at the technology is in order. Like any technology, hybrid vehicles have their advantages and disadvantages, their applications and misapplications.

To understand hybrid technology, we must grasp an elementary physics concept: By virtue of its mass, an object will resist a change in its motion, Newton's first law. Applied to a car, this means that to accelerate or decelerate we must add or extract energy.

When an automobile sets out from a stop, we expend energy to get it going and bring it up to speed. We step on the gas pedal, which sends fuel to the engine, which directs torque to the wheels. A heavier car requires more energy to bring up to speed, a lighter car less. A fast, jackknife start requires more fuel, a slow, smooth start less.

Once at cruising speed, it takes comparatively little energy to maintain that speed, just enough to counteract wind resistance and the drag of engine and drive-train components. A van, SUV, or other square vehicle will have more wind resistance, a sports car less. A pickup, motor home, or eighteen-wheeler will have more engine and drive-train drag, a motorcycle less.

Though most drivers don't think of it in this way, when we slow or stop a vehicle we extract energy from its movement. When we hit the brakes, the car's brake linings push against its brake rotors. This causes friction. The friction transforms the energy of the car's motion, or its kinetic energy, into heat, which is dissipated into the atmosphere. When we slow or stop, we pass to the air the energy contained in the fuel that we used to get the car going.

In this lies the key to hybrid technology. In a hybrid car, when we hit the brakes—at least not in an abrupt stop—the brake linings don't push against the brake rotors and dissipate the car's energy as heat. The brake pedal engages a generator. Turned by the wheels, the generator converts the car's energy of motion into electricity and sends it to a battery. When the car slows, hybrid technology recovers part of the energy that was expended to bring the vehicle up to speed.

It also allows us to use that energy. A generator and an electric motor are essentially the same thing. When the wheels turn the generator's shaft, it sends electricity to the battery. When you take electricity from the battery and direct it to the generator, it works as a motor and the shaft turns the wheels. A hybrid car's generator-electric motor is used in conjunction with the car's gasoline engine to help the vehicle get moving again.

In addition, the car's gasoline engine can directly turn the generator-electric motor or, depending on the hybrid's design, a separate generator and send electricity to the battery. This means that the gasoline engine doesn't have to run all the time but can cycle on and off to assist with acceleration and to maintain a desired charge in the battery. A hybrid car's gasoline engine, for example, will often not run when the car is stopped at a light and thus not expend the fuel to idle.

Hybrid technology has one other advantage. A gasoline engine operates most efficiently within a certain RPM, or revolutions per minute, range. The whole point of gears and a transmission is to keep the engine in this range. Because a hybrid car's gasoline engine works with its generator-electric motor, the gasoline engine doesn't have to be as big and can more often run in its optimal power band.

For the reasons above, hybrid cars have the potential to operate with greater efficiency than gasoline engines. Applicable for the 2008 model year, the EPA applied a new methodology to estimate gas mileage to better reflect real-world performance. Under the traditional methodology, which had been around since the 55-mph speed limit and that used an average highway speed of 49 mph, the 2006 Toyota Prius had city and highway mileage ratings of respectively 60 and 51 miles-per-gallon. Due to a lower top speed, hybrid cars get better gas mileage in the city than on the highway. Under the EPA's new methodology, the 2008 Prius is estimated to have a city and highway mileage of 48 and 45, which is closer to the 37 miles-per-gallon people have been getting. A common complaint among hybrid owners is that their vehicles don't get the mileage posted on the window stickers.

Though real-world hybrid performance is nowhere near that of the traditional EPA estimates, most of us would love it if our car got 37 miles-per-gallon. Hybrid technology clearly has potential. As promising as it is, however, like any technology it has drawbacks.

The car's generator-electric motor and large batteries add weight and size to the vehicle. Batteries also have a limited life and are expensive to construct, replace, and recycle. In addition, the electronic, hydraulic, and mechanical linkages and components needed to coordinate the brakes, wheels, gasoline engine, and generator-electric motor are complex and to a degree new technology.

Components in a hybrid's small gasoline engine will be under more stress than those in a larger, more heavily built engine. In addition, a gasoline engine that cycles on-and-off must make provisions for proper lubrication because most engine wear takes place on startup.

In a cold climate, the gasoline engine may have to run all or nearly all the time to maintain the engine's operating temperature and to warm the batteries and passenger compartment. In a hot climate, the gasoline engine may have to run all or nearly all the time to drive the car's air conditioning compressor. To a degree, high and low outside temperatures negate one of the hybrid's key advantages, the ability to cycle the gas engine on and off.

Thirty-seven miles to the gallon is good mileage, but the Toyota Prius is a small car. The mileage of conventional subcompact cars isn't far behind and can match or even exceed that of a hybrid in some weather and driving conditions. The mileage of SUV and larger hybrids is lower, on the highway often no more than their conventional counterparts. Hybrid or not, mileage depends on vehicle weight, body style, and engine size.

Today, the decision to purchase a hybrid car comes down to the criteria we use—or that at least we should use—to purchase any car or piece of equipment. The owner will be best served if he or she avoids trends and matches use to technology.

If you use your car in true stop-and-go city driving, as one would in say downtown Boston as opposed to on the freeways of Los Angeles where it's more slow and slightly faster driving, hybrid cars have a clear mileage advantage, in particular when it's not hot or cold outside. Over-the-road, the hybrid's extra weight negates the benefit of its smaller engine and it's either a wash or the hybrid falls behind. Hybrid cars also require specially trained mechanics. Whereas a hybrid vehicle might be a good choice if one lives in Dallas where service facilities are around the corner, it might not be a good choice if one lives in the rural west.

When evaluating the purchase of one car over another, it's the difference between vehicles that counts. For most drivers, a small hybrid car at best will give a few more miles-per-gallon than a small conventional car, which even if gas prices increase amounts to a relatively small difference in operating cost. By some calculations, an owner would have to drive more than one hundred thousand miles for fuel savings to offset the premium that at least today is charged to buy a hybrid car. If less stop-and-go driving is assumed, the owner may never recover the cost. It's also important to compare like vehicles—a Toyota Prius with a Honda Civic for example, not a Toyota Prius with a four-wheel-drive Chevrolet Suburban that seats eight and has the horsepower to pull a twelve-thousand pound horse trailer.

Hybrid owners must also be aware of unsubstantiated claims concerning their cars, in particular from companies selling over-capacity replacement batteries with the promise that they'll increase a hybrid car's mileage, in some instances to as much as 200 miles-per-gallon. Oversized battery packs, which typically fill the trunk, allow the owner to plug the car into a wall socket at night and recharge the battery. With greater battery capacity, the hybrid's gasoline engine doesn't have to kick in as soon the next day to maintain the battery's charge. What proponents of the battery scam don't tell you is that it almost always costs more, usually many times the amount, for energy that comes out of a wall socket than for energy in gasoline. They also don't tell you that unless you live next door to a nuclear power plant more fossil fuels will be burned to generate the electricity used by the car than will have been burned had you pulled up to the pump. In the future, hybrid cars that can be plugged into a wall outlet have the potential to be a viable technology. Wide scale use, however, would require a massive increase in the capacity of the nation's electrical grid. Each summer, nearly every major city faces brownouts when temperatures go up and people turn on their air conditioners. What would happen if everyone plugged in their cars?

Prior to Toyota and Honda's production of the first hybrid cars, engineers targeted the technology to commercial delivery vehicles in urban environments. Marketers, however, saw a need for an economical car aimed at the "green" consumer in the United States. Today, you needn't belong to the Sierra Club to consider purchasing a hybrid vehicle. Many new SUV and other models are or will soon be available. If not this year or the next, most of us will face the decision to buy a hybrid. When so confronted, the consumer needs to set aside the emotion of gas prices and global warming and match technology to application. Do you drive in the city or over the highway? Do you need all-wheel-drive, room for a family, the power to haul a boat or trailer? Do you live in a cold or warm part of the country? Hybrid or otherwise, if the product is designed to do what you need it to do buy it, if it isn't buy something that is.

35

Renewable Energy

As of late, the polarization of American politics has reared its obsessive nature with regard to the issue of energy. There are those who call for increasing our nation's energy production through a range of technologies, including coal, nuclear power, and drilling for more oil. There are those who call for conservation and a shift from fossil fuels to renewable sources. What are renewable sources, and will they work?

Like the word "sustainable" so often thrown around by pundits, the word "renewable" as it applies to energy has no clear definition. In theory, it's any fuel or source of power that traps the sun's energy immediately or during a current period of time. Solar collectors, which extract the sun's energy when struck by light, and the production of ethanol from corn, which extracts the sun's energy over a growing season, are said to be renewable. Most often the word refers to any source of energy proponents deem politically or environmentally correct, specifically those that don't involve oil, coal, natural gas, and nuclear power. There are a surprising number of renewable energy sources, and like any form of power generation each has its applications and limitations.

Though technically not renewable, many see geothermal energy as an inexhaustible source of power. And in locations where the earth is warm near its surface, we can extract geothermal energy. The Oregon Institute of Technology heats many of its buildings with water piped beneath the ground to collect heat as does the nearby city of Klamath Falls. The city of Reykjavik in Iceland sits on a vast geothermal resource and uses it to heat buildings, melt snow from roads, and to some extent generate electricity and even produce hydrogen. From a practical standpoint, however, geothermal power is difficult to extract. In most locations, we would simply have to dig too deep to reach it. There are also concerns such as pipe and turbine corrosion and groundwater sources that contain arsenic and other mineral pollutants that when brought to the surface would have to be disposed of.

The nation's largest source of renewable energy is waterpower. The Bonneville, Grand Coulee, and other dams on the Northwest's Snake and Columbia River systems, for example, produce vast amounts of electricity. The aircraft manufacturing industry in Seattle wouldn't be there without them. The Aluminum refining industry in Portland wouldn't be there without them. On the Colorado River, the Hoover Dam that backs up Lake Mead has a generation capacity in excess of 1,300 megawatts. The city of Los Vegas as we know it wouldn't be there without it. But dams have their drawbacks. They impede salmon migration. They also make water available for agriculture and development, which in the minds of some is detrimental to the environment. So vehement is the opposition to dams that there is an active movement to remove them, including many of the largest on the Snake and Columbia.

Next to waterpower, our most important source of renewable energy is the generation of electricity by burning garbage and wood waste. Fueled by private and commercial yard trimmings and by waste products from the timber industry, power generation facilities dot the West. A related source of renewable power is the burning of methane produced as a byproduct of decomposition at landfills and sewage treatment plants. The primary constituent of natural gas, methane is fed into gas lines or burned onsite to generate heat and electricity.

There are also less well known sources of renewable energy such as wave and tidal power. Buoys that move up and down with the swells can drive generators anchored to the ocean floor. Similarly, proposals have been made to dam the San Francisco and other bays and trap water brought in at high tide and drain it through turbines at low tide. The problem with wave and tidal power is that the available energy is dispersed and difficult to harness in quantity. Swells are only so high. To produce a practical amount of electricity, we would have to construct large networks of generators, or offshore energy farms. In a bay, the height of the water-drop, and thus the energy it can convey, is only the few feet of the tide as opposed to the 700 foot drop at Hoover Dam. There are also concerns such as seawater corrosion of machinery and the need for maintenance to remove seaweed and barnacles. In addition to navigation impediments, environmentalists have warned against impediments to fish and whale migration.

Typically, when we think of renewable energy, ethanol, methanol, and biodiesel come to mind.

Much of the gasoline sold in the United States has ethanol mixed in, about ten percent. For older engines in particular, this decreases performance and gas mileage and is corrosive to engine parts. Ethanol, however, is a wonderful fuel when, as we described in our chapter titled *Energy*, an engine is designed to burn it. The problem with ethanol is producing a practical amount of it at a reasonable cost. In the United States, almost all ethanol is made by fermenting corn. When we take into account the energy expended to plant, water, fertilize, harvest, and process corn-ethanol, it takes almost as much energy to make as we get back when we burn it. Sold to legislators under the guise of energy independence, ethanol's use is in fact driven by the agriculture industry and the global warming faction of the environmental community. The *Energy and Independence Security Act of 2007*, for example, provides massive agricultural subsidies and mandates massive increases in corn-ethanol production. Brazil has had some success producing ethanol from sugarcane, which can yield up to eight times more energy output with respect to energy input than corn and most mid-latitude crops. Sugar beets also yield higher energy returns.

There, however, is a way to produce ethanol from sources other than sugar crops—cellulosic ethanol. Ethanol is the stuff we drink and like any alcoholic beverage is made by fermenting sugars. In a crop such as corn, the sugar is in the kernels, and the rest of the plant primarily consists of cellulose, the fibrous stuff in the stalks and leaves. Cellulose is a polysaccharide, a link of simple sugar molecules. To produce cellulosic ethanol, we break down the polysaccharide into its glucose components and then like any sugar ferment them. Though we've been able to do this in the lab for a long time, the technology to do it on an industrial scale is still in the works. The key limiting factor is that it takes a lot of energy to break down cellulose, limiting the net energy extracted.

We, however, can make another type of alcohol from cellulose, one that works even better as a fuel, methanol, or simple wood alcohol. Indy race cars have been burning methanol since the 1960s, and who, on Saturday afternoon television, hasn't watched the fire-spewing methanol burning drag cars thundering down the strip. A number of anabolic bacteria produce methanol from the breakdown of wood and crop waste. Commercially, it's most often made from natural gas. Methanol has one other important characteristic, one that will almost certainly secure its role in our future. It can be used to synthesize more complex fuels: diesel and gasoline.

As for biodiesel, the image of a guy with a ponytail concocting it in his garage from barrels of used cooking oil comes to mind. Eco-friendly communities abound with jokes about retrofitted Volvos rumbling down the street spewing a trail of french-fry smelling smoke. In a simple process that involves the addition of methanol and a catalyst such as sodium hydroxide or sodium methylate, just about any vegetable oil can be turned into biodiesel, which as we mentioned in our chapter titled *Energy* can be used in place of or blended with mineral diesel. Commercially, biodiesel is most often made from soybean oil. Corn, rapeseed, and other oils are also used, and in tropical regions palm oil is used. In 2006, the United States produced upward of 250 million gallons of biodiesel.

Cellulosic ethanol may have potential; and, though we don't hear much about it, methanol will certainly become important. Biodiesel and traditionally fermented ethanol have a drawback that we can only hope will limit their use. They're largely made from or in place of food crops. In 2007, ethanol production gobbled up twenty-five percent of the American corn crop and, when the 2007 legislative mandates fully apply, will by some estimates consume the entire United States crop. Faced with soaring feed costs and pressure from the livestock industry, in 2008 Texas Governor Rick Perry petitioned the EPA to grant a fifty percent waiver on the year's nine billion gallon corn-based renewable fuel standard. To capitalize on the ethanol boom, farmers have moved agricultural land to corn production at the expense of other crops. In 2007, the cost of wheat soared 120 percent, and the cost of rice more than doubled. Global food shortages and soaring costs have led to food riots, most notably in Haiti, Egypt, and Bangladesh. Given population trends, can we live with ethanol mandates and subsidies?

The former oil executive, T. Boone Pickens calls for a shift to wind energy, and he'll be happy to sell you a windmill. Similarly, Al Gore and those who advocate for the global warming issue call for a massive government program to replace fossil fuels with wind and solar power. Advocates, however, fail to take into account the physics of solar and wind power generation and don't grasp, or choose not to grasp, the larger picture of electricity production and consumption.

No matter our politics, we can never harness more solar energy from a square yard of the earth's surface than is made available by the light, or photons, that strike that surface. When solar collector efficiency, energy lost in transmission lines, and factors such as clouds, season, latitude, and nighttime are taken into consideration, to meet the electrical needs of a city the size of Los Angeles, we would have to cover an area the size of Nevada with solar cells. Given this limitation, there is one potentially practical application of solar power: disperse energy collection. In this scheme, solar cells on the roofs of homes and buildings produce electricity during the day and feed it into the grid, allowing conventional power plants to reduce output, effectively storing energy in the grid for use at night and during peak periods. The high cost of solar cells and the reluctance of homeowners to climb on the roof and clean them, however, have limited this method's use. There are technologies on the horizon that increase solar collector efficiency and reduce costs. All factors considered, however, solar energy faces clear physical limits.

At about a tenth the cost of solar power, wind power is more promising. In 2007, the United States produced approximately 16.8 megawatts of power from wind turbines, a fraction of that produced by Hoover Dam but enough to power about one percent of households. Texas is the largest producer of wind energy, followed by California, Iowa, and Minnesota. The current generation of wind generators are mounted on towers as high as 300 feet and are powered by blades as long as 130 feet. And, of course, there are limitations to wind power. Even in the windiest locations, the wind doesn't always blow. The wind can also blow too hard. When winds are above a safe limit, generators must be stopped and the blades secured to prevent damage. In northern regions, ice buildup on blades is a problem. Wind power's most vocal detractors are in the environmental community. In 2003, Senator Ted Kennedy stopped a proposed wind project off the Cape Cod coast, in sight of the Senator's summer home. In 2004, the Center for Biological Diversity filed a lawsuit against the Altamont Pass Wind Resource Area near San Francisco for killing tens of thousands of birds. In 2007, environmentalists filed lawsuits to block wind farms in Texas, expressing concerns over habitat, wetlands, migratory birds, and endangered species.

Solar and wind power have one other significant limitation. No matter its source, electricity produced must equal electricity consumed. Though to some extent energy fluctuations average out, when you turn on a light, in theory an electrical plant somewhere on the grid must step up production to meet the load. When the wind speeds up or slows down or when the clouds break or roll overhead, a conventional power plant must within seconds compensate. For wind and solar power to work, redundant conventional capacity must always be available—boilers up to steam and generators ready to kick in. So difficult is it to compensate for power fluctuations caused by wind and solar sources that in 2002 Denmark placed a moratorium on wind generation, followed by Ireland in 2003.

Renewable sources of power are sometimes referred to as "alternative" sources of energy. There's a reason for this. We don't rely more heavily on renewable sources because of a right-wing, oil-company conspiracy against their use. Most present significant engineering limitations. When practical as opposed to ideological factors are taken into consideration, and the construction of major hydroelectric dams is taken out of the equation, renewable energy sources can at best meet a few percent of the nation's electrical generation requirements. Though politicians may be in denial of any limit to their power, the laws of physics supersede the laws of government.

36

Hydrogen

WITH REGARD TO A clean, reliable, and economical energy source, the United Stated and the world's industrialized nations are in an uncertain position. Hybrid cars or not, and wishful and unrealistic thinking aside, our nation's economy and our way of life are at present contingent on a stable flow of oil. This presents us with a choice. Either we do what it takes to maintain the flow of oil or we do what it takes to develop a replacement for fossil fuels. For the most part, we've chosen the first option. If our military and diplomatic efforts in the Middle East and elsewhere work and we replace theocracies and dictatorships with democracies and respect for individual freedom it will be for our good and for the good of those in affected nations. If we fail, we will have no choice but to as rapidly as possible end our fossil fuel dependence. In the minds of some, this means the use of hydrogen.

The issue of where we get the hydrogen aside, hydrogen technology, like any energy technology, has two basic components: *energy storage* and *energy conversion*, or the transforming of stored energy into some form of useful work.

Hydrogen is the lightest of all elements and in its molecular form, H_2, is gaseous at ordinary pressures and temperatures. These characteristics make storage difficult. So little hydrogen can be compressed into a cylinder that directly compressing hydrogen isn't a practical storage method. A car would have to carry a large, explosive fuel tank. Another solution to hydrogen storage is liquefaction. But unlike propane, which liquefies just under the freezing point of water, hydrogen liquefies just above absolute zero, at about -253 degrees Celsius, or -487 degrees Fahrenheit. Liquid hydrogen at normal temperatures therefore must be kept under extreme pressure. A car with a liquid hydrogen tank would be even more explosive than one with a pressured hydrogen gas tank—a bomb waiting to go off in an accident.

Compression and liquefaction, however, aren't the only ways to store hydrogen. As we know, hydrogen bonds to form hydrocarbons. It also bonds to metals to form a *hydride*. We make a hydride storage tank by enclosing a porous metal lattice, in effect a metal sponge, typically made of a nickel alloy or an alloy of iron and titanium, inside an airtight tank. Hydrogen fed into the tank bonds to the metal to form a hydride and is extracted by controlling the tank's pressure and temperature. For a given volume, more hydrogen can be stored as a hydride than in a standard pressurized cylinder. Hydrogen stored as a hydride is also safe, less volatile than propane, gasoline, or liquefied natural gas. There are, of course, drawbacks. The alloys used to make the metal lattice are heavy and tend to break down. To fill a hydride tank on a warm day, the tank must be cooled or the hydrogen forced in at high pressure. To extract hydrogen on a cold day the tank must be heated or the hydrogen pumped out.

With a way to store hydrogen, we need a way to use it. At its simplest, we draw the hydrogen out of a pressurized cylinder or, if we're more sophisticated, pump it out of a hydride tank, pipe it to an internal combustion engine, and burn it. For years, hobbyists and researchers have been generating hydrogen with lab and backyard electrolysis rigs and fueling cars with modified injectors and carburetors. Because hydrogen isn't available at fueling stations, these cars are usually designed to also burn propane or gasoline.

Burning hydrogen directly is easy to do but not efficient. In an internal combustion engine, much of the chemical energy stored in the fuel is wasted as heat. To increase efficiency, we need to convert hydrogen's chemical energy into a more usable form—electricity. Electricity powers just about everything in our homes and has many advantages in automotive propulsion. Electric engines don't waste fuel idling while the car is stopped at a light; and, as in today's hybrid cars, electric engines can work as generators to recapture energy lost while braking. In an *in-wheel configuration*—or one where an engine is built into each wheel rather than under the hood—it's possible to eliminate axles, drivelines, the transmission, and a single large engine. This reduces the vehicle's cost, weight, and complexity. The problem with electric cars is electricity storage. Despite decades of work, scientists haven't been able to develop a practical battery, one that charges quickly and stores enough energy to provide a reasonable driving range.

Hydrogen eliminates this need. As far back as the early 1800s, experimenters made *fuel cells* that converted hydrogen's chemical energy directly into electricity. A fuel cell performs electrolysis in reverse. In electrolysis, we feed electricity through an electrolytic, or conductive medium, water with a little salt added, which breaks into hydrogen and oxygen. In a fuel cell, we feed hydrogen and oxygen past electrodes separated by an electrolytic medium in which the hydrogen and oxygen combine into water and produce electricity. A fuel cell car spews steam and a spattering of water out the tailpipe. How icy will the roads be in the winter? In today's cars, engine efficiency runs at between 20 and 30 percent for gasoline engines and as high as 40 percent for diesel engines. In on-the-road use, where factors such as wind drag and time spent sitting at stoplights are taken into consideration, today's cars convert on the average about 10 percent of gasoline's energy and about 13 percent of diesel's energy into usable work. By some estimates, an electric-hydrogen fuel cell car can in on-the-road use convert 40 percent of hydrogen's chemical energy into work, though a figure of between 20 and 30 percent is more realistic.

As intriguing and potentially useful as fuel cell technology may be, it has one significant limitation. Using a typical electrolytic medium such as potassium hydroxide, fuel cells only work at high temperatures. Low-temperature fuel cells have been developed that utilize a catalyst to increase the rate of the hydrogen-oxygen reaction. Catalysts such as the metals platinum and palladium, however, are rare and expensive. Will we swap oil wars in the Middle East for Palladium wars in central Africa? To be widely implemented in automotive technology, fuel cells need better electrolytic mediums and cheaper, more available catalysts.

Another significant obstacle to hydrogen's use is distribution. Pipelines from production facilities to fueling stations and homes are one possibility. But the hydrogen molecule is so small that it can easily escape from pipes, making it difficult to, as some have proposed, transport through existing natural gas lines. Another option is to run electric lines to on-site electrolysis facilities, to gas stations or even home units that produce hydrogen.

The dream of environmentalists to have a hydrogen fuel cell car pull up to a solar-powered hydrogen generation station, however, is just that, a dream. How many acres of solar panels would it take to fill a hydride tank as fast as a gas tank, and how many more acres would be needed to fill car after car, at night, in the rain? Contrary to widespread opinion, the slow adoption of hydrogen hasn't been due to a government, oil industry conspiracy against its use. General Motors alone employs more than 500 engineers in hydrogen propulsion development. For hydrogen to be practical, we need hydride storage tanks that are lighter and last longer, fuel cells that work better in a car or truck, and an infrastructure to distribute the hydrogen.

Though at present hydrogen has limitations as a transportation fuel, it has another application—one that is not often talked about in the media but that may position hydrogen as central to the world's energy future. Hydrogen can combine with carbon to form methanol, or simple wood alcohol. In this process, carbon dioxide, CO_2, extracted from the atmosphere, recovered as a byproduct of combustion, or even recovered as a byproduct of converting coal to diesel and gasoline is broken down to release carbon monoxide, CO. The carbon monoxide then reacts with the hydrogen to form the methanol, CH_3OH. As we mentioned in our chapter titled *Alternative Energy*, methanol can then be used as a transportation fuel and as the base to synthesize complex hydrocarbon fuels. The difficulty with methanol and synthetic fuels is that it takes more energy to make them than you get back when you burn them. In economics terms, they cost a lot. If, however, you had an abundant source of cheap, electrical energy, you could at least in theory produce as much fuel of just about any composition that you wanted and eliminate any need for oil and coal extraction or for the use of agricultural land to produce fuel crops. Moreover, such would be a scalable, purely industrial process. Imagine a factory that with no input other than an electrical utility line pumped out diesel, alcohol, gasoline, and jet fuel. Hydrogen, not as we think of it in terms of fuel cells and hydrogen powered cars, but as a component in the chemical reactions needed to synthesize higher level fuels may be our future.

This brings us to the next chapter and to the technologically most proven but politically most controversial aspect of hydrogen technology and the fuels that hydrogen can be used to manufacture. To generate hydrogen on a scale where it can play any significant role in our energy future means one thing—nuclear power.

37

Nuclear Power

WHEN WE CONSIDER THE world's energy situation and the strengths and weaknesses of energy technologies, we come to an unavoidable conclusion. If our goal is to scale back the use of oil, coal, and other fossil fuels—a worthwhile endeavor for many reasons—solar, bio-fuels, and other renewable energy sources can't fill the gap. The numbers don't add up. By the most optimistic estimates, renewable sources can at best provide a few percent of our energy needs. This leaves the world and the United States with two options. We shrink the global economy—end development in India, China, and the third world and accept massive standard of living reductions in Europe, the United States, and other industrial regions—or we turn to the one truly abundant source of energy we have available—nuclear power. One can even argue that the endorsement of global warming in some scientific circles is nothing more than a tactic to maneuver the environmental community into the acceptance of nuclear power. The evidence for humanity's contribution to climate change is, arguably, weak. But environmentalists buy into it, and nuclear power solves the problem of greenhouse gas emissions whether or not it exists. Given the hard facts of the energy situation, it's only a matter of time before the United States reverses its seventies-era antinuclear policies and implements a massive deployment of nuclear power generation facilities. For this reason, we need a commonsense understanding of the issue.

As you may recall from your high school or college physics courses, an atom has a nucleus made of protons and neutrons. Nuclear fission takes place when the nucleus of a complex atom breaks apart to form the nuclei of less complex atoms. In the process of this *decay*, energy is released—a lot of it, up to ten million times as much as in a chemical reaction.

In the area of power generation, the most common reaction is the decay of the uranium-235 isotope. The reaction is initiated when a neutron strikes the nucleus of the uranium-235 atom and results in the release of energy and the production of Cesium-140, Rubidium-93, and three additional neutrons. These neutrons then go on to strike other uranium-235 atoms, which causes a chain-reaction and a continuous release of energy.

To regulate this chain reaction, a reactor is designed to control the flow of neutrons. This is accomplished by separating uranium fuel rods with a *moderating medium* and by removing or inserting neutron-absorbing materials between fuel rods. Carbon, sodium, hydrogen, deuterium, and other substances have been used as moderating mediums. In the United States, and in other developed nations most reactors use highly purified water.

These *light water reactors* are tried technology and have many advantages. But they have a drawback that in today's world is important. They require a more refined fuel than natural uranium. Natural uranium contains only a small amount of uranium-235, 0.71 percent, with the rest made up of non-fissile uranium-238. To be used in a light water reactor, natural uranium must be *enriched*, or refined to between 3.0 and 10.0 percent uranium-235, depending on the reactor's design. The problem with this is that uranium enrichment facilities can also be used to create the uranium-235 ratios necessary to work in a weapon. This is the concern when on the news we hear that gas centrifuges and other uranium enrichment machinery have been detected in Iran, North Korea, and other nations that may want nuclear power for reasons other than to toast bread.

Light water reactors also have a functional disadvantage. The decay of uranium-235 produces a remarkable amount of energy in comparison with the burning of a fossil fuel but, from a nuclear standpoint in a light water reactor, is inefficient. A typically nuclear plant will only harness about one percent of the energy content of the uranium—and uranium-235 deposits aren't inexhaustible. By some estimates, United States reserves could only provide thirty percent of electrical power needs for the next fifty years. Other estimates are considerably higher.

But, although uranium-235 is comparatively rare, it can be used to produce an abundance of other nuclear fuels. This takes place in what is popularly referred to as a *breeder reactor*. When excess neutrons are absorbed by a *fertile* material that material is changed, or transmuted, into a fissile material, typically uranium-238 into plutonium-239. In this reaction, the plutonium then absorbs a neutron and breaks apart, which releases energy and neutrons and produces more plutonium.

To maintain this reaction, a breeder reactor typically uses a molten metal as its moderating fluid, often liquid sodium which melts just under the boiling point of water. A breeder reactor plant can operate as high as 75 percent efficiency while producing 20 percent more fuel than it consumes. Nuclear researchers envision a future where there are various types and sizes of nuclear power facilities based on standard designs, similar to the French model of today. Ideally such an implementation would balance the type and capacity of power plants to produce as much waste, or fuel, as they consume. For the sake of illustration, however, if the United States did nothing more than replace its one hundred or so existing light water power plants with breeder facilities, it could eliminate the use of all oil, gas, and coal for building heating, electricity production, and other stationary uses. This would free up natural gas for liquefaction and use as a transportation fuel, open the door to hydrogen production, and end any conceivable need to import oil. The first power-generation breeder facility was built in France in 1984. Others have been built in Russia and Great Britain. Today, so much uranium-238 is recoverable from waste and the enrichment process that existing stockpiles could support a full-scale deployment of breeder reactors for centuries.

Nuclear technology has been with us for a long time. It's well understood, and there's no question that it can provide as much energy as we could use well into the future. Conceivably, the initiation, or even the announcement, of a program in the United States to eliminate fossil fuel use for electricity production and other non-transportation uses would cause such an uproar in the energy markets that the cost of diesel, gasoline, and jet fuel would plummet. But there's the issue that makes nuclear power controversial—safety.

In part, this is because nuclear power plants need and create materials that can be used in nuclear weapons. When the United States and the Soviet Union were the dominant nuclear powers, the possibility of engagement was perhaps not as great as we may have been led to believe. These nations understood the power of their weapons and, one could argue, knew better than to use them. The factions bent on acquiring nuclear weaponry today, however, didn't share in its development and, judged by the way nuclear threats are thrown around by some leaders, may have no real understanding of what such weapons can do. Case in point, the Islamic terrorist who in 2007 was caught smuggling weapons-grade uranium in his pocket. As important, these factions often have no geographical base and as such aren't susceptible to the threat of nuclear retaliation. Whom would the United States launch against?

The most persuasive safety concern, though, is the possibility of an accident. Concrete walls shield reactor cores. Plants have backup cooling systems and house reactors in concrete containment buildings. But even with redundant safety features, accidents have happened. In 1979, the Three Mile Island plant in Pennsylvania lost its cooling system. When the backup system failed, the core overheated and dumped radioactive materials into the containment building and a small amount into the atmosphere. In 1986 at Chernobyl in the Soviet Union, one of four early-technology reactors that lacked any form of containment building partially melted down, contaminating the surrounding area and spreading radioactive material over northern Europe.

The United Nations estimates that approximately one hundred people died or will eventually die as a result of the Chernobyl disaster, those in the initial explosion and those as a result of cancer caused by radiation exposure. On the other hand, in Europe and the United States, no one has died or been injured in a radiation related nuclear power plant accident and no plant, including Three Mile Island, has emitted radioactive materials of any consequence into the water or atmosphere. A few scientists did lose their lives in the pioneering days of nuclear research, but the modern nuclear energy industry has a safety record that surpasses that of any other industrial enterprise. In contrast, thousands die every year in oilfield and coalmining accidents, tens of thousands die from fossil fuel related air pollution, and hundreds of thousands die as a result of energy related political unrest.

A nuclear power plant is also seen as a target for a terrorist attack. The reactor in a nuclear power plant, the part that contains the radioactive material, though, is small and easily hardened. Containment buildings are heavily reinforced, and even those built decades ago were designed to withstand bombs and plane crashes. If a cooling tower or other exposed equipment is hit, the plant simply shuts down until repairs are made. So resistant to terrorism are nuclear facilities that, behind the scenes, security experts hope that if a terrorist targets anything it's a nuclear facility. The damage and loss of life would be far less than for other targets.

Also misunderstood is the issue of waste. Spent fuel from light water reactors contains almost all the original uranium-238 and about one-third of the original uranium-235, which can be recycled. The main reason waste has accumulated is that, due to political factors, the United States hasn't developed breeder and reprocessing programs. By some estimates, a family of four using breeder generated electricity over a period of twenty years would generate no more than a few grams of nuclear waste, a tablespoon or two. The actual danger of nuclear waste has also been overstated. Most of the radioactivity in waste is produced by a few hot isotopes that have short half-lives. Typical nuclear waste will "cool" to the point where it can be easily handled and poses little danger in about fifty years. In France, reprocessing facilities let nuclear waste sit for about five years before working with the material. The present solution to waste is onsite storage at nuclear facilities, though attempts have been made to create centralized storage facilities in stable geological formations, Yucca Mountain in Nevada. In reality, the label of nuclear waste and even the need for a central storage facility are tools of the antinuclear movement. California, for example, has banned construction of all new nuclear facilities until a permanent waste disposal site is in place. We've long known how to safely dispose of nuclear waste in undersea calcium formations, where new calcium deposits continuously form and would entomb waste for hundreds of millions of years. We don't dispose of nuclear waste because nuclear waste isn't waste. It's one of our nation's most valuable energy resources.

Different nations perceive safety differently, and technology has advanced from the Soviet and Three Mile Island era. Sweden and Austria have limited or terminated their nuclear programs. In the United States, no new nuclear plants have been ordered since 1978; and, at an estimated cost of one hundred billion dollars, many completed plants haven't been allowed to operate. On the other hand, Japan, France, Germany, and Great Britain have active nuclear programs but face stiff environmental opposition. Current instability in the Middle East and North Africa, however, has encouraged nations to reconsider the nuclear power option. Canada has plants in the works, and China's continued economic growth hinges on the massive deployment of nuclear power. Even in the United States, utilities have submitted applications to construct new nuclear facilities. In the current political climate, however, it's estimated that it will take approximately ten years for companies to exhaust the legal and legislative hurdles environmental groups have vowed to present.

In the minds of some, the threat to our way of life posed by the oil producing nations and their grip on our energy supply is as great as that posed by Adolf Hitler during World War II. What would life be like in New York, Los Angeles, or any city if oil supplies were cut off? To confront this situation, we must look past the trivial energy policies of the past. No more gas taxes, luxury taxes, and mileage standards. No more passing bills that encapsulate politics in ineffective and unenforceable mandates. Solar, bio-fuels, and wind power have their applications but cannot to any significant extent meet our needs. Game over. We have one feasible energy option—nuclear power.

38

Energy Plan for America

Every day pundits and politicians talk about foreign policy, economic policy, and environmental policy. Underlying these issues is a larger policy issue, one that at this point in history virtually every aspect of governance hinges on—energy policy. Economic activity may rest on the creativity and initiative of the individual; but, for the individual to express his or her creativity and initiative, he or she must have the ability to reshape and interact with the material world—and that requires energy. The United States exerts its influence around the globe; but, as George W. Bush's Secretary of State, Condoleezza Rice, has said, the need to maintain the flow of foreign oil severely limits America's foreign policy options. Human caused global warming may or may not exist, but no matter our environmental politics it doesn't make sense to burn vast amounts of oil, coal, and natural gas and dump into the atmosphere the pollutants these fuels contain. One of the measures anthropologists use to gauge the level of a culture's development is the amount of energy each member of a society consumes. As over time new cultures emerged and humanity progressed, this number has increased. One can argue that energy consumption isn't the only measure of cultural development and that as societies evolve and adopt new urban designs it may not be valid at all. At present and for the decades immediately ahead, however, America's progress and the progress of humanity are contingent on the generation of vast new amounts of energy.

This point established, we turn to the matter of America's energy policy. When one faces a problem and embraces logic and common sense as the tool to solve it, a scenario of three steps emerges. First, we begin by asking ourselves what we know about the situation. What knowledge tools do we have at our disposal?

As we established, America and the world will not only need to accommodate our current energy demands but develop vast new sources of energy. We can't conserve our way to energy independence. This doesn't mean that we shouldn't build cars, homes, and light bulbs that use less energy, but that conservation isn't the answer or even a significant part of the answer. To impact our use of energy through conservation, we would have to end development in the third world and shrink the economies of the developed nations—thrust the world into economic depression. Though we may be able to use less energy in certain situations, we must produce more.

We also know that from the American standpoint, we must eliminate our dependence on foreign oil and other sources of imported energy. As a nation, as the world's leading economic and military power, we can't allow ourselves to be held in the grip of foreign interests, nor channel our wealth to these interests. This isn't best for us. This isn't best for the world. Moreover, given the uncertainties of the global economy, we must eliminate our dependence on foreign energy as rapidly as possible. Not a century from now, but as soon as from a technological standpoint we can make it happen, we must secure America's energy independence.

Our energy policy must also take into consideration the practical limitations of energy technologies. We have no economic or moral right to divert America's agricultural production to fuel crops. This isn't fair to the American consumer who must pay for it in higher grocery prices. This isn't fair to those around the world who depend on America's agricultural production for subsistence. Similarly, we must accept the limitations of so-called alternative energy sources. Biomass and hydroelectric power have proven potential. Wind and solar power have engineering limitations that no matter the lobbying efforts of their supporters we can't get around.

Energy production is a matter of physics and engineering. Politics has and will continue to enter into our energy policy, but to successfully solve the nation's energy problems, science and reason must trump corporate, environmental, and any other interest that wishes to impose their agenda over workable solutions.

This brings us to the second step in our logic and common sense derived scenario for solving the nation's energy problem. After we've identified what we know, we must identify where we want to be. What is our objective? From an energy standpoint, what kind of a world do we want to create?

Clearly, we want an abundance of energy and for that energy to be available to us from a variety of fuels and other sources. Top on this list would be electricity. Ultimately, we must eliminate the use of fossil fuels for stationary power generation. As soon as we can, we must get rid of our oil, gas, and coal fired electric plants. Hydrocarbon fuels contain a lot of energy per unit of mass and volume. For this reason, they are too valuable as transportation fuels to use in stationary applications where portability isn't a significant factor. With regard to transportation fuels, we must make available different formulations for different applications: diesel, alcohol, and gasoline for automotive and heavy equipment use, jet fuel and high octane gasoline for aviation use.

When we take into consideration all engineering factors, one source of power stands out as essential to our energy future—nuclear fission. When we objectively look at all our energy options, nuclear fission is the only technology that from any practical standpoint has the potential to meet America and the world's energy needs. This doesn't mean that we abandon alternative sources or the quest for nuclear fusion. Wind and solar power can contribute, and fusion may one day become a reality. Based on the technologies available in the foreseeable future, however, human progress is contingent on a massive, wide-scale deployment of nuclear fission. But we must expand our use of nuclear power safely. We must control and monitor reactor byproducts. We must harden reactors against terrorist attacks. We must design reactors that aren't susceptible to meltdown.

As important, we must adopt a coherent strategy to deploy nuclear power. Plants need to be located in a way that minimizes the need for electrical power transmission lines. Electrical transmission lines waste tremendous amounts of energy and in certain areas leave huge scars on the landscape. In addition, we need a blend of nuclear technologies—conventional and breeder facilities of the type and number required to draw on America's stockpile of nuclear byproducts, or so called waste, with the goal to produce the amount and type of nuclear fuel in the amount and type we consume, effectively eliminating all concerns over waste storage.

When we look at our energy future in its entirety, we see a world that functions through a nuclear infrastructure of the capacity to meet our electrical energy needs and the capacity to, as natural sources of fossil fuels become more scarce and expensive, generate the hydrogen and provide the source of power to break down carbon dioxide into carbon monoxide and synthesize methanol. With methanol as a base and with a source of power, we can synthesize diesel, gasoline, and jet fuel. Nuclear power also opens the door to other energy intensive applications, the wide-scale desalination of seawater for example, critical in many of the world's coastal regions.

In our problem solving scenario, we've identified what we know and where we want to be. The final step is to determine how we will get there. How will we achieve our objective to create the energy future we envision?

Above all, we must have a plan. Where do we want to be six months from now, one year from now, ten years from now? Our plan must be detailed and rigorous, but it must also be flexible. We will learn as we go. Above all, our plan must eliminate political barriers. The model to do this is legislation passed in the 1970s that blocked legal challenges to the construction of the Alaska Pipeline. Legislators recognized the importance of this pipeline and felt that it could be built safely. They also knew that the legal challenges the environmental community threatened to pose would tie up the project for decades. Across the energy spectrum, whether it be nuclear power or offshore drilling, it isn't technology that lends uncertainty to our energy future—it's politics. To implement a workable energy future, we must remove the legal and political obstacles to its implementation.

Now to establish the specifics of our energy plan. Within a reasonable timeframe, say a ten year period, we need to move all, or the vast majority, of stationary power generation to a nuclear infrastructure. This will entail several things: First, as we mentioned, we need to determine the correct balance of nuclear designs and the optimal placement of nuclear plants. Second, we must develop standard designs for power plants, mass produce reactors and generating facilities. Third, coinciding with nuclear plant placement, we must reconstruct the nation's electrical grid to meet our demand for electricity and to deliver it to our homes and industries in the most efficient and least environmentally damaging way. Fourth, we must bolster math and science curriculums at the K-through-12 and college levels, produce the scientists and engineers to build and operate our nuclear infrastructure.

With regard to our nation's electrical grid, the reader must be wary of what is often referred to as the "smart" grid, promoted by environmentalists and by corporations who will profit from its construction. In the smart grid, our home and business electrical meters, and conceivably every appliance, would via a chip communicate with the utility company. Proponents claim that this will save energy. Whether computer controlled or not, a toaster still needs a certain amount of energy to toast a slice of bread. What the so called smart-grid actually does is allow utility companies to charge different rates for different users at different times of the day. To reduce demand and the need for fossil fuels and generating capacity, utilities would increase the price of electricity during peak periods.

As we put nuclear plants on-line, we open the way for restructuring our energy use and production in other areas. With what by some assessments will be nuclear generated electricity available at a fraction of a cent per kilowatt hour, we need to move home and business heating and hot water generation from oil and gas to electricity. This, and the phasing out of fossil fuel use in power generation, will free up tremendous amounts of traditional fuel. Ultimately, our goal must be to move our automotive and other transportation systems from fossil fuels to hydrogen and, more realistically, to methanol and other synthesized fuels. Until the electrical generation capacity and the infrastructure to accomplish this become available, however, we must take intermediary steps. Conversion of coal to diesel and gasoline is one option. Electric cars and plug-in hybrid cars is another option. Most practical of all, we can move areas of our transportation system to liquefied natural gas. This would be particular easy to do for trucks and other fleet vehicles where we can readily put into place the fuel delivery infrastructure.

As inspiring as our energy future may be, we must also embrace the energy realities of the present. For the first several years, until our nuclear infrastructure is in place, our transportation system will continue to run on diesel, gasoline, and jet fuel. In the short term, we must drill for more oil—Alaska, offshore, wherever we can find it. We, of course, must extract it with a commonsense level of environmental risk. Critics claim that it would take years to bring offshore and other new fields into operation. This is only true when one takes into account the legal hurdles critics vow to pose. It took just over two years to build the Alaskan Pipeline. By some assessments we could have new sources of offshore and North Slope oil flowing within months. Critics also point out that oil pumped in the United States doesn't stay in the United States; it enters a global market. What we do with the oil we pump in our shores is up to us. Almost all the oil we presently extract out of Alaska ends up on the West Coast. Because of a lack of refining capacity, we ship oil to Japan to be refined and then ship it back. But for the most part, the oil we pump out of Alaska and elsewhere in the United States stays in the United States.

Our energy plan will have long and short term components, but it must be coherent—an exercise in engineering creativity and not the product of political bickering by scientifically illiterate factions. As a nation, we must also make a commitment to our energy plan. As the recent plunge in the price of crude has shown us, the global cost of oil is highly sensitive to demand. When Ronald Reagan set the goal of energy independence in the 1980s and put into place a synthetic fuels initiative to achieve it, the Saudis ramped up production. For a time during the 1990s, crude fell to less than ten dollars a barrel, which ended the political will for energy independence. No matter the short-term nuances of the market and the maneuvering by parties who profit from the status quo, we must stick with our energy plan, govern for the long term, move forward with an unflinching eye on the future.

To establish our nation's energy policy, we must evaluate what we know, envision where we want to be, and draft a working plan to achieve our objectives. The rebuilding of America's energy infrastructure is a grand undertaking, and government can't do it alone. Government may function as the driving force and as an overall coordinating mechanism, but to establish the nuclear and synthetic fuels infrastructure of tomorrow we must have a public-private partnership as formidable as in any war effort. But we can do it. We have no choice but to do it. When we look beyond politics, we no longer see America's energy situation as a problem. We see it as an opportunity. The energy predicament of today offers us as a nation the means to reinvigorate our inventiveness and lead the world into an age of energy abundance, and the peace and prosperity energy abundance will bring.

39

The Middle East

IN TERMS OF ENERGY, terrorism, the nation of Israel, military involvement in Iraq and Afghanistan, and a myriad of other events and factors, life in the United States is on nearly every level connected to life in the Middle East. To begin to understand this most troubled region of the world and its peoples—and to lend perspective to the chapters on Islam and terrorism that follow—we need a brief history.

The modern story of the Middle East takes us to the fifteenth century when tribes in the Anatolian region of what is now Turkey united to form one of history's most notable reigns—that of the Turks, or *Ottoman Empire*. From tribal beginnings and the 1453 capture of the Christian capital of Constantinople, the Turks spread their influence throughout the region and within seventy years had united the Middle East and penetrated as far as central Europe. Not since the last remnants of the Eastern Roman Empire had the region been under a single rule—one that would last for the next 400 years and only end after the First World War.

But the Ottoman reign was not without strife. By 1700, France, England, and other European nations had grown wealthy and powerful. With the technological advantages gained during the industrial revolution, they drove the Ottomans out of western and central Europe. In the nineteenth century, Britain and France occupied parts of North Africa and the Persian Gulf. In response, the Ottomans allied with Germany and, as a consequence, became dependent on that state for financial and geopolitical security. In the early twentieth century, Turkish ties to Germany strengthened, and Middle Eastern leaders setout to modernize the Ottoman realm. Often referred to as "Young Turks," these new leaders built railroads, universities, and communication lines and aspired to establish secular civil codes, constitutional governments, and public education systems. This modernization trend, however, was threatened by the power and influence of traditional Islamic leaders and scholars. As such, the movement never gave way to the stability of true democratic regimes. Desperate to stay in power, the leadership of the Young Turks abandoned idealism, reverted to authoritarianism, and led a series of brutal military juntas.

The attempt to modernize and spread Ottoman cultural values also fueled a rise in Arab nationalism and dug the empire into debt with Germany. This alienated other potential European partners and with the onset of World War I, positioned the empire to side with Germany. As a weak link in the Axis, the Ottoman Empire became a strategic target of the British, who sought to establish alliances with nations unhappy with Ottoman rule. At the end of the war, the allies partitioned the Ottoman territories through a number of agreements. These culminated with the 1920 Treaty of Sèvres and the 1923 Treaty of Lausanne, which allowed Constantinople, now under the name of Istanbul, to remain in Ottoman hands, created an Armenian state, and spread control of the Arab provinces and the remaining regions between Britain, France, Greece, and Italy.

European control also set the stage for events that would carry the Middle East through the world's second great conflict.

During World War I, Britain had allied with a number of tribal and regional leaders. Among these men was Sherif Hussein who led a successful Arab revolt against the Ottomans in the Arabian Peninsula in exchange for the promise of future independence. The promise of independence, however, was never fulfilled, and in the *Sykes-Picot Agreement*, Britain split the region with France. Arab leaders felt further betrayed by the British *Belflour Declaration*, which supported the establishment of a Jewish homeland in the Arab dominated region of Palestine.

To ease tensions and minimize Arab resentment after Britain withdrew its backing of Arab independence, France, which had established protectorates in Muslim dominated Syria and Christian dominated Lebanon, established two kingdoms for Sherif Hussein's sons: *Iraq* for Faisal and *Transjordan* for Abdullah. In the western half of Palestine, Britain allowed the Jewish population to increase under its protection and, to appease another wartime ally, *Ibn Saud*, in 1922 established the Kingdom of Saudi Arabia.

Soon after in Persia, or Iran, a wartime Guerilla fighter, *Reza Khan*, overthrew what remained of the long-reigning Qajar Dynasty and declared himself Shah, or king. Reza Khan initiated industrialization, railroad construction, and the establishment of a national education system. But he also maintained ties with Germany that when World War II broke out alarmed Britain and the Soviet Union. In 1941, Britain and Russia invaded Iran and forced Reza Shah to abdicate in favor of his son, Mohammed Reza Pahlavi, known in the West as the *Shah of Iran*.

The division of the Middle East into kingdoms was also influenced by another factor, one of growing importance throughout the world but in particular in Western Europe—oil. Discovered in Persia in 1908, in Saudi Arabia in 1938, and in other Gulf States soon after, Middle Eastern oil was essential to Western Europe's industrial growth. It also made Middle Eastern kings and emirs profoundly rich—the wealth they needed to maintain their hold on power and stifle progress by Middle Eastern nations toward democracy and modernization.

While Britain and France pursued their interests in the region, Turkey continued on the path of modernization it had begun decades earlier under the Young Turks. It distanced itself from its cultural and geopolitical relationships with the Arab world and drew closer to the West. Throughout the Second World War, the Middle East was a battleground for oil and, at the war's end, was once again under Western control. From agreements at the end of the war, however, the region would emerge a much different place. Iran, Iraq, Syria, Egypt, Jordan, Lebanon, and Saudi Arabia became independent states. And, in 1947, the United Nations initiated a plan to partition Palestine into two nations—one Jewish along the Mediterranean Sea and one Arab along the Jordan River.

Jewish leaders agreed to this arrangement; Arab leaders did not. Immediately following the 1948 establishment of the State of Israel, the armies of Iraq, Syria, Egypt, Jordan, Lebanon, and Saudi Arabia attacked Israel, to be convincingly defeated by a newly formed Israeli army. In the aftermath, nearly 900 thousand Arabs fled Israel to become what for the most part were unwelcome refugees in neighboring Arab states—a situation that over the next forty years would help to incite a number of major conflicts between Israel and its neighbors.

Among these was the 1967 *Six-Day War*, during which in a preemptive strike considered one of history's most skillful military actions Israel defeated the combined armies of Egypt, Jordan, and Syria backed by the military support of Iraq, Sudan, Tunisia, Algeria, Morocco, and Saudi Arabia. As an outcome, Israel gained control of the Gaza Strip, West

Bank, Golan Heights, East Jerusalem, and Sinai Peninsula. The world gained continued Arab-Israeli tensions and an increase in terrorism. This included the 1972 Munich Olympic Massacre where Islamic extremists killed eleven Israeli athletes and a number of others at the Munich games in Germany.

Also as a consequence of the formation of an Israeli state after the Second World War, the United States superseded Europe as the dominant external influence in the region; and, as tensions between America and the Soviet Union heated up, Arab nations became pawns in the greater struggle of the Cold War. Among Middle Eastern countries, Nasser in Egypt and Saddam Hussein in Iraq adopted strategies of "nonalignment" with the United States and established relations with the Soviet Union. To counter, the United States shored up relationships with regimes in Jordan, Saudi Arabia, and in particular in Iran; where, in 1951, *Mohammad Mosaddeq*, a long-time adversary of the Shah's father, was, with the backdoor encouragement of the Soviet Union, elected the country's Prime Minister.

Mosaddeq nationalized Britain's decades of investment in Iran's oil infrastructure, a move that made him popular at home but not in Britain, who with America's cooperation orchestrated a plot to depose Mosaddeq that led to his arrest in 1953 and solidified the Shah's hold on power. With American support, the Shah launched a program to modernize the Iranian military and the nation's infrastructure. Iran's military presence had a stabilizing influence on the region that lasted decades, but the Shah's openness with the West inflamed fundamentalist Moslem elements. In 1978, American president Jimmy Carter orchestrated the Camp David Accord, which established peace between Egypt and Israel that has lasted to this day. American policy with regard to Iran showed less foresight. That same year, fundamentalists revolted, overran the American Embassy, and took more than one hundred Americans and others hostages. This, followed by the decline in health and ultimate death of the Shah, led to the return from exile in France and the rise to power of Iran's most revered and fanatical religious leader, the *Ayatollah Ruhollah Khomeini*.

The Iranian Revolution also ended American aid and led to a near collapse of the Iranian military. In neighboring Iraq, Saddam Hussein took advantage of instability in the region and in 1980 launched an invasion of Iran, an action that initiated the *Iran-Iraq War*. Not militarily strong enough to counter Saddam Hussein's advances in a conventional way, the Ayatollah recruited tens of thousands of young boys to, in a tactic of Martyrdom, storm Iraqi forces until the enemy had depleted its ammunition and could be overrun by the regular military. So effective was this tactic that after driving back Hussein, the Ayatollah proclaimed the goal to move through the Middle East and unify it as a fundamentalist Islamic state. Faced with the possibility of a regional conflict that would threaten the State of Israel and Middle Eastern oil supplies, the United States and a number of Western and Middle Eastern allies countered by backing Hussein and funded a massive buildup of the Iraqi army. With a staggering death toll on both sides, the use of chemical weapons, and battles reminiscent of the trench warfare of World War I, the Iran-Iraq war raged for eight years. In 1988, the Ayatollah overreached and blocked oil shipments in the Persian Gulf. United States warships responded by sinking the bulk of the Iranian navy—a move that brought Iran to the table and ultimately brought the war to an end with a United Nations Security Council Resolution.

The Iran-Iraq war also left Saddam Hussein in command of a powerful Iraqi army. This, along with the collapse of the Soviet Union in 1991, thrust the Middle East into its most recent historical period, one that would draw Europe and the United States into the region to an extent not seen since World War II.

In 1979, The Soviet Union had launched an invasion of Afghanistan to stifle resistance by the loosely aligned and predominantly fundamentalist *Mujahideen* resistance, or "freedom fighters," against the ruling Marxist government. To counter, the United States under Carter and Reagan—along with Pakistan, Saudi Arabia, and the Europeans—backed the Mujahideen. The war in Afghanistan raged for more than nine years until, faced with economic difficulties in the Soviet Union, the Soviets withdrew in 1988. This left a civil war between the government and the Mujahideen, which as the government weakened led to a struggle within the resistance that played out by the rise to power of the fundamentalist *Taliban* regime.

All the while, in 1990, Saddam Hussein in Iraq, on the pretext of a territorial dispute over oilfields, invaded its southern neighbor Kuwait. The brutality of the invasion stunned the Western world. Hussein's aggressiveness also brought terror into the hearts of the Saudis, who believed that the Iraqi dictator would make good on his rhetoric to move through the Middle East and unite the region under Bagdad rule. Fearful of Israel's security, fearful of a disruption in the flow of Middle Eastern oil, and under intense pressure from the Saudis, United States President George Herbert Walker Bush organized a coalition of 500 thousand American, European, and Arab troops and in the First Gulf War, or operation "Desert Storm," drove Hussein out of Kuwait.

After the war, Bush initiated economic sanctions that, though the cost was primarily bore by Iraqi citizens, kept a grip on the country for more than a decade. The first Gulf War also increased America's presence in the region, a move that in part led to further radicalization of fundamentalist elements. Violence increased between Israel and its neighbors, and for the first time to any significant extent terrorist tactics spread outside of Europe and the Middle East. Among fundamentalist groups was *Al-Qaeda*, led by the son of a wealthy Saudi family who rose to power as a Western-backed fighter against the Soviets in Afghanistan—*Osama bin Laden*.

Through his Al-Qaeda organization, Osama bin Laden is believed to have orchestrated the 1993 World Trade Center bombing in New York, the 1998 bombing of the United States embassies in Kenya and Tanzania, and the 2000 bombing of the USS Cole in Yemen. Al-Qaeda also orchestrated numerous bombings in Egypt, Yemen, and Saudi Arabia. Bin Laden and Al-Qaeda's mission of terrorism reached an apogee with the September 11, 2001, hijacking of American airliners that resulted in the strike on the Pentagon and the destruction of the World Trade Center. These events, in turn, bring us up to date in that they incited United States involvement in the Middle East such that, in terms of success and failure, not enough time has passed for history to judge—the invasion of Afghanistan in 2001 and the invasion of Iraq in 2003.

40

God

No matter how hard we work to maintain the "cherished" barrier between state and church—God and religion find their way into almost every issue. To understand this and to benefit from a situation that we face every day at home, work, and school, we need to contrast God with religion and let this take us where it may.

When most of us think of God, we think of religion. We don't pray to a nameless or faceless god. We pray to a well-known and regularly worshiped god, one that's given form by religious tradition—to Buda, Allah, Christ, Brahman or some other familiar image of God. To most of us, God and religion are the same, or at least inseparable. But when we look at God and religion closer to their theological roots, they appear as two different things.

As Eastern, Native American, and other religious traditions that emphasize spiritual awareness suggest, God has to do with experience. God is the sensation we feel, or believe we feel, that we're connected to a higher consciousness, that in some way we're linked to a loftier plane of existence. God is the sense that we're embraced by a universal awareness, that we're part of something that is ourselves and that at the same time is greater than ourselves.

Archeological evidence of hunting and funeral rites tells us that our experience of God, or of some form of supreme being or beings, dates remarkably far back in time. For at least the last one hundred thousand years, we've pondered our sense of connectivity to a higher consciousness and sought to appeal to that consciousness for a better life and for insight into how and for what purpose we came to be—the nature and meaning of our existence.

In contrast, religion isn't something we feel or experience; it's something we think. A religion is a set of ideas and concepts we create and share that are intended to bring us closer to the experience of God and to help us come to terms with that experience. A religion puts a face on God and offers explanations for the universal human questions of meaning and purpose. A religion is a system of beliefs we invented and propagate that allows us to understand, or in some way to think we understand, God and the human condition.

At least this is religion in the lofty sense. A system of beliefs can be about anything. Some see our devotion to capitalism and free markets or, conversely, to socialism and government control as a religion. Earlier, we referred to environmentalism as a religion. God is god. It's something we experience or believe we experience. As the anthropologist defines it, a religion is a *shared system of beliefs*.

It's difficult to say if our experience of "God" has changed over time, but it's clear that we've reinvented our beliefs about God and about the human place in existence countless times.

To our earliest ancestors, mystical beings controlled the migration of game, the passing of day into night, birth and death, and the mishaps and moments of good fortune that take place in every person's life. Over time, our images of and relation to these beings refined and grew more complex. The ancient Greeks envisioned gods that controlled the sky, sea, and earth from their home on Mount Olympus—Zeus, Hera, Ares, Athena, Apollo, Artemis, Aphrodite. Each city had its favorite gods and honored them with prayers, sacrifices, and celebrations.

Thirty-five hundred years ago, the Egyptian ruler Akhenaton, established the belief that the "sun god" was the all-embracing spirit of the universe, and *polytheism*, the notion of multiple gods that controlled different aspects of existence, was challenged by *monotheism*, the notion of a single god that controlled all aspects of existence. The early Jews embraced monotheism, but felt that it was all right for other nations to have other gods. Christianity, which emerged from Judaism, was less tolerant. The early Christians felt that their god was the only true God, which put them at odds with the Romans who worshiped a plethora of deities.

As Christianity spread, our notion of God solidified. Today, every major religion embraces a remarkably similar conception of God. To nearly all religious practitioners, God, or with regard to certain eastern philosophies a godhead, is the creator, an ultimate reality beyond the human power to fully understand. God stands above us in judgment and is to be worshiped, even feared. God is an all-powerful, all-knowing, ever-present being.

This conception of God creates certain dilemmas. Foremost is the concept of *good and evil*. If God is an all-powerful, all-knowing, ever-present entity, why would God unleash on humanity the suffering and ignorance it has for so long endured? Would not an all-powerful, all-knowing, ever-present god see to it that we lived in a perfect world created by its perfect hand? Why does God not simply tell us the answers to life's most important questions?

To account for evil in the world, a religion must adopt certain explanatory notions. Foremost among these is the concept of *freewill*. In one way or another, every major religion embraces the notion that the human being has the autonomy to conduct his or her affairs with or without God's approval. Though in most cases, God's desires are somewhat hard to determine, and our lives are judged by the norms of religious propriety.

Also important is the idea of an *evil force*. To some degree, every major religion attributes human suffering and ignorance to the action of a force or being whose will opposes that of God—Satan, Lilith, Asmodeus, or other personification of evil or foe of good—demons versus angels. A person responsible for committing evil is said to be aligned with the entities of evil. The victim of evil is said to have drifted from his faith in God.

The problem with the ideas of freewill and evil force is that they challenge the power and wisdom of an all-powerful, all-knowing, ever-present God and thus the foundation on which every major religious philosophy is based. This is a problem that theologians have long grappled with and that one would think in time would lead to a new conceptualization of God. On these lines, philosophers have speculated on the existence of an evolving God, or a God who grows and learns through the tribulations of the human experience. On the other hand, a growing number of Americans embrace ideas of spirituality not associated with a God. We find meaning in nature, family, helping others, and leading a good life. Problems with the traditional view of God also support ideas of atheism, naturalism, and existentialism, or the belief that God doesn't exist and that all answers are found by way of reason and observation of the natural world. At what point, however, does the belief that God doesn't exist become a shared system of beliefs—a religion? The man or woman who denies the existence of God as part of his or her personal view of existence, or who is agnostic on the issue, is nothing more. He or she embraces a personal interpretation of existence. Is the activist who demands that the word "God" be taken out of the United States Pledge of Allegiance fighting for the barrier between state and church or to impose his or her "godless" religion?

No matter what our religion or lack thereof, religious ideas have and will continue to shape our values and judgment. God is experiential, something we feel, believe we feel, or believe we can feel and as such take to exist. Religion is a shared system of beliefs that at its highest nurtures our experience of God and addresses the fundamental human questions of meaning and purpose. To prevent itself from dissipating into a sea of philosophies, a religion must set itself apart from other ideologies. It demands adherence to doctrine and tradition, often at the expense of furthering the experience of God the founders of that doctrine and tradition may have envisioned it to achieve. To embrace the distinction between God and religion is to free oneself from the bounds of dogma and open oneself to the possibility of religion at its most inspiring.

41

Islam

ON THE TOPIC OF religion, one faith has in recent years dominated our thoughts—Islam. In light of global terrorism, the popular media's analysis of Islam and the contemporary study of the faith have given us two schools of thought. One addresses the religion without judgment. Islam is a faith no better or worse than any other, and acts of evil committed in its name are the responsibility of those who perform them. The other sees the Islamic faith as in part responsible for, or at least condoning of, certain acts of evil. In this chapter, we provide the history and background needed to follow and perhaps make sense of these arguments.

As we described in our chapter titled *Civilization IV*, the Islamic faith began in the early seventh century with Muhammad's vision of the biblical archangel Gabriel. Revealed to Muhammad in this vision, Gabriel explained that throughout the ages God had selected prophets on earth to teach nations and individuals moral and spiritual behavior. Prophets represent the best of humanity; and, since they are the conduits through which God's word is conveyed, the religions they found are much the same. Here, though, Muhammad's vision stands apart from those of earlier prophets. In the words of the archangel, Muhammad would be given a perfect revelation of God that superseded all before. As such, he would become the last prophet.

At first, Muhammad confided his vision only to his family and friends but in time began to preach in his native city of Mecca. Ridiculed by the Meccans, in 622 he fled to Medina. This event, the *Hegira*, marks the beginning of the Islamic calendar. At the time, religion on the Arabian Peninsula was largely polytheistic: a mix of Greek, Roman, and Egyptian gods with Jewish and Christian traditions thrown in. In Medina, the climate for religious change was right, and the "last prophet's" message of the universal, all-powerful God revealed to him by Gabriel was well received.

Muhammad's followers established him as a spiritual leader and as one of Medina's political leaders. His base of power in place, Muhammad crushed Arab, Jewish, and Christian opposition in the city and in command of its army declared war against Mecca. In 630, Mecca surrendered. The Region's Bedouin tribes offered their allegiance; and, by the time of his death in 632, Muhammad commanded an Arab state poised to exert its power and influence.

And between the seventh and twelfth centuries, it did. Arabian armies spread Islam throughout the Middle East, Asia Minor, and North Africa. In 751, the center of Islamic power shifted from Mecca to Baghdad; and, in the centuries that followed, Islamic armies out of Egypt and Turkey projected Muslim rule into the Spanish Peninsula and turned back the Crusades.

As military propagation unfolded, Islamic scholars established Islam's fundamental laws and theology. Like other major religions, at the heart of Islam is a conceptualization of God, or *Allah*, and of humanity's relationship to the divine.

In the Muslim faith, God created the earth and cosmos from nothingness as an act of mercy. Each element of creation has its place and behaves according to immutable moral and natural laws. This results in a harmonious whole, a universe of orderly function presided over by God and that, by way of order and function, serves as proof of God's hand in its governance.

The purpose of the human being in this universe is to serve God. This is done when one devotes his or her life to "reforming the earth," which means to establish the universe's proper and natural order by bringing any who may disrupt it, all who aren't Muslim, into the service of Allah.

On the *Day of Judgment*, those successful in this task, and who have performed the deeds to prove it, will go to the *Garden*, or heaven. Those corrupted by power, pride, and wealth will go to hell. Like individuals, nations succumb to the evil of power, pride, and wealth and, in the Muslim act of reforming the earth, are to be destroyed or subjugated by virtuous nations.

This ideal of the divine and of humanity's servitude to the divine is contained in the Islamic faith's two primary sources of doctrine and practice: the *Koran* and the *Sunna*.

Muslims regard the Koran as the word of God proclaimed to the world through Muhammad. As such, God is taken to be the author of the

Koran and its words and meaning to be infallible. Tangibly, the Koran is divided into 114 chapters, each of which contains as many as 306 verses. Unlike the New and Old Testaments, which are generally believed to have evolved over their existence, scholars think that do to its comparative youth and early recording in written form today's Koran has maintained much of its original wording.

The Sunna, or the example of the prophet, represents the traditions established by Muhammad. These are contained in a book called the *Hadith*. Unlike the Koran, which as we mentioned was compiled in written form quite early, the Hadith was passed orally between the generations until it was put on paper in the ninth century. Because the Hadith was told by men, based on the life of a man, and had been passed verbally for almost two centuries, it's not considered infallible. Many Muslims, however, consider it to be as important as the Koran.

The Islamic ideal of God and the beliefs and practices embodied in the Koran and Hadith are given form by the five major duties, or *Five Pillars*, of Islam. The Muslim's first duty is faith: to at least once in his or her lifetime publicly profess his or her conviction: "There is no God but Allah and Muhammad is his Prophet." The Muslim's second duty is prayer: In response to a mosque's call to congregate, the faithful are to five times a day bow to Mecca and pray. The Muslim's third duty is *Almsgiving*: or the payment of *zakat*, a tax originally levied by Muhammad and though voluntary today is seen as essential. The Muslim's fourth duty is the *fast*: During the month of *Ramadan*, he or she is not to eat, drink, smoke, and have sex from dawn until sunset. The Muslim's fifth duty is *pilgrimage*: At least once in a lifetime, all who are physically able are to make the pilgrimage to Mecca.

As with other faiths, throughout history there have been different interpretations of Islam. The first major theological school was established by the *Mutazilites*. Influenced by Greek philosophical works translated into Arabic in the eighth and ninth centuries, Mutazilite scholars felt that moral truths could be established through logic and reason. In the tenth century, this view was challenged by the orthodox, or *Sunnite*, theological school. Its scholars felt that moral truth could only be established by God and known to man through revelation.

Between the ninth and twelfth centuries, attempts to reconcile Greek ideals of freewill and rationality with Sunnite ideals of absolute faith and literal interpretation of the Koran led to a mystical movement called *Sufism,* which through moral purification was devoted to the direct communion with God. This threatened Muhammad's claim to be the last prophet, and the movement was put down by the Sunnites and disregarded by most educated elite.

Another sect, the *Shiites* emerged out of a dispute over political succession. Claimed to be the descendents of Muhammad through his daughter Fatima and her husband Ali, the Shiites took, and continue to take, the rule of Islam to be their divine right. Until the ninth century, this right was executed by a line of twelve "infallible" leaders. The last of these rulers mysteriously disappeared in 880; and, as the Christians await the return of Christ, the Shiites await the return of the Twelfth, and the world of order, justice, and morality he is promised to bring.

In the eighteenth century, the Syrian jurist Muhammad ibn Abd al-Wahhab founded the puritanical *Wahhabi* sect. The Wahhabis rejected the principle of consensus, or *Ijma,* in scriptural interpretation as well as all forms of music, dance, tobacco, and gambling. In the nineteenth century, the Wahhabis successfully attacked the Islamic shrine at Karbala and the cities of Mecca, Riyadh, and Medina but faced defeat against the Turkish sultan Mahmud II.

Today, approximately eighty million of the world's more than one billion Muslims follow various interpretations of Shiite Islam. With the exception of about eight million Wahhabis in the Arabian Peninsula and a handful of minor sects mostly in Asia and North Africa, the remainder follows various interpretations of orthodox Sunnite Islam. Derived from the Koran, Hadith, and centuries of practice, Islamic values and behavior are spelled out by *Islamic Law,* and Allah's divine service of reforming the earth is implemented through the *Jihad.* Translated to mean "holy war," Jihad may take place militarily but may also take place through the democratic or other subjugation of the political institutions needed to implement Islam throughout the world.

In many ways, Islam is a religion like any other. All major religions have their fundamentalist interpretations. All embrace a remarkably similar conception of God and deal with the philosophical dilemmas it creates in much the same way. All set themselves above other faiths, and all seek to propagate their vision of God and the universe. But the Islamic faith of today isn't the Jewish or Christian faith of today. Throughout Islam's history, orthodox forces have harshly put down attempts at modernization. Judged by the absoluteness with which many Muslims interpret Islamic doctrine and practice, the Islam of today is in certain respects like the Islam of the seventh century. Is not the struggle that wages in the Islamic world and that filters into the West, the turmoil of reformation? Is it not the battle that has always waged between the novel and the status-quo, the fundamentalist ideals of the past and the free and individualistic ideals of the future? Perhaps it is this understanding that will allow the world to come to terms with Islam and for those who find meaning in Muhammad's teachings to enter into and prosper in a religiously diverse, secularly governed twenty-first century.

42

Terrorism

SINCE BEFORE RECORDED HISTORY, one dimension of who we are has more than any other pervaded the human experience—violent conflict. Violent conflict revels itself as murder and warfare. It reveals itself as the building of nations and empires, and as their decline and collapse. It also reveals itself as one of the most difficult to understand and deal with characteristics of modern life—terrorism. We face religious terrorism. We face economic terrorism. We face environmental terrorism. We face men and woman who don't value life as we do and that further their agendas with methods we weren't brought up to think possible.

To come to at least some understanding of this most perplexing of topics, we need to step back. We begin our look at terrorism with the rudiments of human conflict.

In the nonhuman world, violent interaction between members of a species establishes an animal's place in a social group, and violent interaction between social groups establishes a group's place in the larger set of species connections that define the ecosystem. Wolves fight among themselves for dominance in the pack. Packs defend their territory against neighboring packs.

In the early hunting and gathering society, violent conflict unfolded in much the same way. We fought for social position in our group and for our group's place in an ecosystem, defined by its territory. Anthropologists tell us that this type of aggression was common but not to the extent we might think. Social structure was uniform and collective and social status was comparatively easy to establish. The ecological arrangement of groups was also clearly defined. As long as a group respected its neighbor's territory there was little to fight about.

As social structure rose from collectivity, we acted in increasingly violent ways. We vied for power and social position and aligned to dominate neighboring groups by military means and to fend off the military advances of others. With urbanization, our political needs entwined with our material needs. We battled for power and status and for land and material possession.

We also fought for ideology. What war is not waged in the name of a God or Supreme Being? How many campaigns have we engaged in over an economic or political ideal? Ideology allows us to justify our aggression. We fight for the greater good we feel our economic and religious beliefs will bring to our families and to those whom we conquer.

The Prussian military theorist *Karl Von Clausewitz* described war as an extension of *Politics*, which we described as the tactics and strategy of social rearrangement. However we may choose to express violent conflict, it achieves political ends through forcible means.[1]

In the context of this definition, we see terrorism as a way to exert force, as a tactic or method of warfare—but one that manifests in only certain situations. Terrorism takes place when the opposing ideology, government, or institution is militarily too strong to take on directly.

For more than a decade, Ted Kacznski, the Unabomber, waged war against the advance of technology by mailing letter bombs to scientists and industry leaders. For more than two decades, the Earth Liberation Front has burned ski lodges, sport utility vehicles, housing complexes, and timber industry plants and offices. In 1993, the Animal Liberation Front planted incendiary devices at Chicago area department stores to stop the selling of furs. That same year, Islamic terrorists bombed the World Trade Center in New York City. And in 1995, Timothy McVeigh bombed the federal building in Oklahoma City.

The Russian Revolution of 1917 was preceded by a generation of terrorist acts, which culminated in riots and in the popular support needed to overthrow Tsar Nicholas II. In 1979, Iranian terrorists overran the United States Embassy in Tehran and took 103 hostages, holding them for more than a year. In 1983, suicide bombers killed almost 300 French and United States Peacekeepers in Beirut, Lebanon. In 1988, Libyan terrorists downed a Pan Am Boeing 747 over Lockerbie, Scotland. Since its formation by the United Nations in 1948 and its occupation of the Gaza Strip and West Bank in the Six-Day war of 1967, Israel has faced wave

---

1. In the broader philosophical sense, we see violent human conflict as one way in which social evolution has manifested, as one mechanism through which the human rise out of collectivity has taken place.

after wave of Palestinian terrorist attacks, in recent years conducted by suicide bombers.

Rarely are terrorist acts performed by a lone individual, a Ted Kacznski. Almost always, they're conducted by an organization fighting for a political cause, and some agencies define terrorism strictly in this way. A terrorist organization begins with, or is transformed by, a person who for a personal reason, the thirst for power, or for a benevolent reason, the perceived betterment of society, is driven to bring about political change by terrorist tactics.

But for this leader to transform his ambitions into reality, he needs to enlist the allegiance of others. This can only take place in a social context where people are looking for someone to lead them to a better future. Inflation and national humiliation in Germany after World War I fueled Hitler's rise to power. Throughout the Middle East, unemployment, perceived injustice, and perhaps at the core a lack of freedom and opportunity drive men and women into the ranks of terrorist groups. Conversely, feelings of need or inadequacy among the affluent leave students and others vulnerable to the promises of activists and cult figures.

To recruit from the disenfranchised, a leader needs one other ingredient—ideology. Hitler focused his ambitions with a message of Jewish hatred and world domination. Islamic terrorists focus their ambitions with a message of religious fundamentalism and hatred of Western wealth, morality, and liberal ideals. In 1993, the Branch Davidian leader David Koresh focused his dream of religious leadership with a message of death and Armageddon.

To combat terrorism, therefore, we must deal with it on each of these levels. In today's world, there will always be someone driven by the thirst for power or someone who for some perceived good wants to bring about political change and sees terrorism as a path to this end. To stop suicide bombers, we not only have to intercept the bombers but those who made and handed them the bombs and those who talked them into sacrificing their lives for a cause.

But when we eliminate one leader, another will invariably fill the void. We also have to deal with the conditions that allow a leader's ambitions to manifest in a violent way. Those bent on world domination are nothing more than a Ted Kacznski locked away in a cabin if they can't draw others to their cause. Underneath stated objectives, questions concerning the accuracy of intelligence and the motivation of the Bush Administration, and the politically unmentionable but inescapable variable of oil, the United States invasion of Afghanistan in 2001 and of Iraq in 2003 had a purpose that went beyond reducing the ranks of those who would do the world harm. At least in part, we invaded to instill, or begin the process of instilling, open economies and governments in the Middle East with the hope that we could improve life in the region to the point where the average person had no reason to take up a call to arms. Historians will judge the success of our efforts.

Last, we must deal with whatever ideology a leader has co-opted to further his or her agenda. Environmental, animal-rights, and similar groups base their ideology on a politically selective interpretation of science. The Northern Spotted Owl nests in "old-growth" forests in the United States Pacific Northwest. To save the "endangered" spotted owl, environmentalists in the 1990s lobbied to ban logging in these forests and politicians eager for campaign contributions from the "green" community stopped almost all timber harvest on public lands. Today—after laws were passed, logging curtailed, an industry devastated, and thousands of family-wage jobs lost—the northern spotted owl is threatened by the common barred owl, which is better adapted to the dense, overgrown forests our good but misguided intentions created. The antidote to politically selective science is science and an understanding of what science is and how it works.

When it comes to religious and in particular Islamic ideology, however, the strategy isn't as straight forward. Religion is embedded at nearly every level of Islamic society. To combat Islamic terrorism, the West must abandon a popularly held belief—the notion of *cultural equality*, also referred to as cultural or moral equivalence. Preached into the mindset by anthropologists in the 1960s, it's the idea that every culture is of equal value. It's the belief that no society is better or worse than any other only different, the notion that a society in the Solomon Islands that one hundred years ago raised young girls to be sold into kai-kai, or to be eaten, is as deserving of respect as one that today sends rovers to Mars. At a 2007 protest in Washington DC against the Iraq war, activists encapsulated this view in their rhetoric. Speech after speech was given predicated on the belief that problems in the Middle East were the result of America's lack of respect for and understanding of the Islamic world.

Our lack of respect for the Islamic world may be unfounded, but our lack of understanding of the Islamic world is a reasonable assessment. Few in America, or anyone outside the Islamic world, can justifiably claim to understand that world. But to blame conflict in the Middle East—a region that has been in upheaval since the birth of civilization—on America—or more accurately, on capitalistic greed and Western imperialism—is to deny the fact that cultures change over time. It's to reject history, to disregard millennia of cultural evolution.

It goes without saying that every human being has intrinsic value. All men and women are born equal and have the potential for growth and betterment. But societies advance at different rates. A society in the Solomon Islands that one hundred years ago raised young girls to be eaten is more primitive than one that today sends rovers to Mars. With respect to Islam, we don't stone rape victims. We don't execute homosexuals. We don't hack off the hands of shoplifters. We don't block women without headscarves from fleeing a burning building. We may tolerate a president who fools around in the oval office, but not one with a dozen wives and fifty children.

And most residents of the Middle East are as repulsed by these acts and the values they represent as you and I are. As we concluded in our chapter titled *Islam*, conflict in the Islamic world is a battle between the past and the future, between values held and practices performed in the seventh century and values held and practices performed in the twenty-first century. The Islamic world is embroiled in a period of transformation that in the view of many analysts is in certain respects analogous to that experienced by the Catholic Church during the time of the Inquisition and Martin Luther's protestant reformation.

Rooted in this conclusion is another observation, one leaders in and out of the Islamic world would do well to keep in mind. As evidenced by Catholic reformation, cultural evolution reflects an uplifting of consciousness, the embracing by the individual of a new understanding of the world. To be enduring, therefore, such cannot be imposed from the outside or from the top down but must come from within. Conflict in the Middle East can be resolved, but ultimately only by the citizens of the Middle East.

But such is not the justification for pacifism that the antiwar activist may take it to be. By virtue of the industrial world's need for oil and immigrants, the increasing interconnectedness of the world, and radical Middle Eastern regimes aspiring to be armed with nuclear weapons—the West is caught in the middle of Islam's struggle, embroiled in the throes of Islam's transformation. The border of nearly every Middle Eastern nation was at one time drawn by a Western hand. The bloody Iran-Iraq war ended in 1988 when the United States navy decimated the Iranian navy and thwarted the Ayatollah Khomeini's plan to move into Egypt and Saudi Arabia and take control of the Middle East. Two years later, Iraq's Saddam Hussein invaded Kuwait, and a United States led coalition of 500 thousand troops decimated the Iraqi army and thwarted Hussein's plan to move into Egypt and Saudi Arabia and take control of the Middle East. We in the West don't have the option to bury our heads in the sand of cultural equality. We may wish to deny our place in the world and our status in cultural evolution, but circumstance compels us to play a role in Islam's reformation.

We, however, must do so with deliberation. We must confront terrorist leaders, end recruitment of the disenfranchised, and challenge terrorist ideology. At times, diplomacy, economic persuasion, and legal and police measures are best. At other times, military force, which common sense tells us must be taken to be the option of last resort, is the only answer. Gripped by ideology to such an extent that, as we saw in the Japanese kamikaze raids of World War II and see in the suicide bombers of today, we can't always succeed with the give-and-take of negotiation. Without General Douglass McArthur's iron hand at the end of the Second World War, would Japan be the open society we know today? No matter our tactics, however, we must direct our efforts to establishing the conditions where change can unfold from within. The West's role isn't to impose the twenty-first century on Islam. Its role is to, directed by common sense and noble intentions, help create the social and economic environment where those in the Islamic world who want the future can stand up to and overcome those who stagnate in the past.

43

Darwin

As we described in our chapter titled *God*, a religion is a shared system of beliefs. This system of beliefs may be about a Supreme Being and the meaning of the universe, but it may also be about others things. Sometimes it even bears the name of science. One such system of beliefs, the theory of *Natural Selection,* was proposed 150 years ago by *Charles Darwin*. Today, no other theoretically construct is more deeply embedded in science and in modern culture. And no other theoretical construct is more misunderstood.

To better appreciate Darwin and what he did and didn't contribute to the world, we need to contrast evolution with the theory of natural selection.

Just as most of us see God and religion as the same or inseparable, most of us see organic evolution and Darwin's theory of natural selection as the same, or inseparable. When we talk about organic evolution, we talk about natural selection. When we talk about natural selection, we talk about organic evolution. But organic evolution and natural selection aren't the same.

Evolution is an event, something that took place. It's the map of organic transformation, the sequence in which life forms came into existence. The event of organic evolution is supported by a vast body of fossil evidence, hundreds of millions of finds that to a remarkable extent encompass the entirety of life's existence on earth. The fossil record may not be complete in every respect, but it exists to such an extent that no reasonable mind can deny its message: New forms of life emerged from old forms of life. Paleontologists may not be able to tell us every nuance of this emergence, but they have gathered such evidence as to make it incontrovertible.

In contrast, natural selection is a theory that attempts to explain how the observed and well documented event of organic evolution took place. In natural selection, mutations, or changes, are said to randomly occur in organism DNA and to create variations in organism structure and behavior. If these variations strengthen the organism, and it's better able to survive its environment and reproduce, it passes its improved DNA to its offspring. Naturally occurring genetic traits are "selected" by the environment—natural selection.

On one hand, we have evolution. On the other, we have Darwinian natural selection, a theory that attempts to explain how the observed phenomena of evolution took place. In no other way are the two related.

To better understand this distinction and to carry our discussion forward to its consequences, it would be useful to learn a little about Darwin and how he came up with his model.

Born in Shrewsbury, England, in 1808, Darwin briefly studied medicine at the University of Edinburgh but left medical school to complete his education at the University of Cambridge where he planned to become a clergyman. Rather than pursue a career in the church, he sailed as an unpaid naturalist aboard the HMS Beagle on a scientific expedition around the world.

Aboard the Beagle, Darwin noticed that fossil remains of extinct species closely resembled the anatomical structure of living species. He also observed that members of the same living species in different parts of the world looked somewhat different. Most scientists felt that species were created individually and never changed over time. From his observations, Darwin concluded that species did change and that earlier forms evolved into later forms.

On his return to England, Darwin spent years trying to figure out how this evolution of species could have taken place. At the time, England was an empire in the throes of the industrial revolution. Life was hard and competitive. Men struggled to feed their families. Companies struggled to extend their influence around the globe. Life was dominated by the contradictory views of science and religion and the conflict they incited.

From our vantage in the present what is clear, and most scientists concur, is that in formulating his theory of natural selection, Darwin to some extent imposed his cultural ideals on his observations of the natural world. Darwin framed his observations of species and how they changed over time in the context of what he knew and saw taking place around him—struggle and competition. Evolution was survival of the fittest.

How well, then, does Darwin's theory of natural selection account for evolution? Though we may have been taught in school that Darwinism was a matter of fact, in reality the issue isn't as clear. The event of organic evolution is for all intents-and-purposes universally accepted by the scientific community. Evolution clearly took place. Despite what we may believe, however, Darwin's theory of natural selection is widely held but by no means embraced unequivocally as the process through which that evolution took place.

Chief among the theory's critics was Darwin himself. Darwin saw the natural selection model as a starting point, as a framework that in time would lead to a more explanatory understanding of the evolutionary mechanism. So aware was Darwin of the limitations of the simple adaptive mechanism on which the natural selection model is based that, with regard to the anatomy of the human eye, he felt that it was "absurd" to think that something so complex could be the product of random mutation and environmental testing.

Though Darwin may not have been aware of it at the time, his concern over biological complexity revealed larger problems with the natural selection model. In the fossil record, we observe coherent lines of evolutionary ascent. Primitive forms give rise to successively less primitive forms, a process sometimes referred to as *orthogenesis*. Natural selection however, tells us that evolution has no direction. It progresses randomly. Less complex species can evolve into more complex species, and more complex species can evolve into less complex species—whichever has a survival and reproductive advantage in a given environment. This, however, isn't what we observe in the fossil record where rather than devolve species fall into extinction. Similarly, a species exists as part of the network interactions we refer to as the ecosystem, which exists as part of successively more complex networks of interactions—all which are in evolution. Because natural selection takes place on the level of the species, some scientists feel that a more sweeping evolutionary mechanism must be at work.

Among the theory's critics was the twentieth century French philosopher and paleontologist *Pierre Teilhard de Chardin*. Teilhard, who participated in the "Peking Man" and other major archaeological finds, embraced the event of evolution as indisputable, but felt that a more comprehensive explanation was necessary. Inspired by an earlier philosopher,

Henri Bergson, Teilhard felt that evolution took place through the *creative process*, through the same creative mechanism that we experience in ourselves. This view solved many of natural selection's problems. But because Teilhard died before he could fully develop his theory, it fell into obscurity. Today, Teilhard's *creative evolution* is gaining renewed interest.

When confronting a scientific edifice like natural selection, it's important not to confuse challenges made by those who sincerely want to advance scientific understanding and explain observed behavior with those who want to further an agenda. This steers our discussion to politics and to a topic that has been a subject of controversy for decades and even today occasionally makes the news—the issue of creationism versus evolution.

In the United States, creationists are primarily, though not exclusively, Christian fundamentalists who propagate the belief that man and earth were created as they interpret the Bible or other scripture. Often under the label of "Intelligent Design," the creationist movement lobbies legislators and school boards to require the teaching of the biblical account of creation as an alternative to evolution in high school and college science courses. Aware of problems with the theory of natural selection, creationists argue that it doesn't work and, since evolution is the same as natural selection, evolution didn't take place. Moreover, since evolution didn't take place, the fossil record must be invalid—either misinterpreted or a fabrication of the scientific community.

Conversely, the scientific community counters with the argument that the physical evidence for evolution is overwhelming and, since in the minds of most scientists evolution is the same as natural selection, Darwinism must be the mechanism through which it took place. A surprising number of scientists have read the Bible and other scriptural documents. The problem isn't necessarily that scientists disagree with or wish to challenge another individual's religious beliefs; it's that they don't feel those beliefs should be taught in a science course. Religion is religion and, as we discussed in our chapter titled *Science*, science is science.

This debate has gone on for decades and in that both arguments are predicated on the false assumption that evolution and natural selection are the same thing, will probably go on for some time to come. When we set aside this predication, we temper the creationism versus evolution debate with common sense. Natural selection may not fully explain evolution, but that doesn't mean that evolution didn't take place and as a consequence the only other explanation is that God is responsible. Biblical and other creation accounts are a matter of faith and not of science; and, for this reason, they belong in a religious studies class and not in a biology, physical anthropology, or other natural science course. Darwinian natural selection may also embody an element of faith, but at present it's the model that the majority of scientists embrace to account for evolution and as such belongs in a science course. Scientists, however, have an obligation to make the distinction between evolution and natural selection and to teach Darwinism as theory and not as fact. Many scientific edifices have gone-by-the wayside. Physicists no longer believe that light travels through the ether. Theories of dark matter and dark energy have dramatically changed our view of the cosmos. Darwin's theory of natural selection is the generally accepted explanation for organic evolution, but it may not be the final explanation.

The natural selection view of organic change over time has profoundly impacted Western society. Darwin incorporated his worldview into his theory of natural selection; we then used Darwin's theory of natural selection to justify our worldview. Politics is survival of the fittest. Economics is survival of the fittest. Education is survival of the fittest. Competition creates jobs. Competition improves business. Competition drives down prices. In our "every-man-for-himself" world, it's acceptable to have some who are left behind, for in nature there are winners and losers. The next time you face or make an argument, ask if it's justified by Darwin's ideals of survival and competition and think about his theory of natural selection. The event of organic evolution is a fact supported by a vast body of fossil evidence. If Darwinism proves to be no more than a step along the way in our understanding of the event of evolution, the impact on our view of the world and on our way of life will be profound.

44

Clones

On the subject of God and Darwin, it's impossible to ignore one of the most controversial lines of scientific research now underway—cloning. How does cloning work? What is the difference between therapeutic and reproductive cloning, and what philosophical issue does cloning force science to confront?

The first mammal to be successfully cloned was the famous and short-lived "Dolly" the sheep. Dolly was cloned through a process called the *nuclear-transfer* procedure.

In reproduction, a female egg cell has half a normal cell's genetic material, and the male sperm cell has the other half. In the nuclear-transfer procedure, a technician removes the genetic material from an egg to create a genetically "empty" cell. The full complement of DNA is then extracted from a nonreproductive cell—in the case of Dolly taken from a cell removed from the udder of the animal that would become its mother—and inserted into the egg. This gives the egg the full complement of DNA; it thinks it's fertilized and begins to divide. When this *embryo* reaches an early phase in its development called the *blastocyst* stage, it's implanted into the uterus of a female where, if everything goes right, it will mature to the point of birth. Because no genetic material from reproductive cells is used, the nuclear-transfer procedure results in an exact copy of the animal from which the full complement of DNA was extracted.

Though simple in concept, the nuclear-transfer procedure doesn't always work. Cells have different developmental phases. To remove and extract DNA, the cells have to be at the right phase, and the phase has to be properly arrested and activated. The techniques used to culture cells, transfer DNA, and transfer embryos to a uterus also leave room for error. And there are factors scientists haven't even begun to understand. The animal from which Dolly was cloned was six years old. Dolly died while still young, arguably from old age. When the genetic material from the donor was transferred into the recipient egg, the lifespan clock appeared not to be reset. Even with recent advances, the cloning process is hit-or-miss. A lot of eggs have to go through the nuclear-transfer procedure and be implanted into surrogate mothers to end up with a "normal" birth, in the case of Dolly 277.

Even so, rats, sheep, horses, and monkeys have been cloned. And, with the exception of the animal rights people, nobody much cares if animal cloning doesn't always work. No medical process has been perfected without trial and error. When it comes to humans, the uncertainties of cloning combined with moral, family, and societal issues make cloning a contentious matter. *Reproductive cloning*, or cloning to produce a man or woman identical to a donor, is universally renounced. There are those who for reasons they alone must understand want to be cloned and there are researchers who want to be the first to clone a human being, but no scientist has thus far taken on the legal and moral hurdles to pursue the process to the point of a birth. But there's another kind of cloning, and its ethical issues are less clear—*therapeutic cloning*.

In therapeutic cloning, cells taken from the embryo at the blastocyst stage are allowed to replicate in a cell culture. At such an early stage in the embryo's development, these cells have a remarkable quality. Called *stem cells*, they can divide to become any kind of cell—a nerve cell, a liver cell, a heart cell, a kidney cell. Not only could therapeutic cloning eliminate organ shortages, a replacement organ grown from stem cells cloned from the recipient would have that person's genetic makeup and thus not be rejected and require the use of anti-rejection drugs. Growing a human heart is in the future, but embryonic and in particular adult stem cells, similar cells found in the mature body, have been used to treat diabetes and Parkinson's disease with some promise.

With so much potential to bring good into the world, why is therapeutic cloning controversial? Here our discussion crosses the line from science to philosophy.

In our chapter titled *Self*, we asked what sets the human being apart from the animal and concluded that we're reflective beings, that we have the capacity to ponder the nature and meaning of our existence. In therapeutic cloning, controversy centers on the question of at what point in our fetal development do we acquire the capacity, or at least the clear potential, for self-awareness. At what point are we a mass of cells and at what point are we a human being or unequivocally on the road to becoming a human being?

Just about everyone agrees that at the end stages of pregnancy the fetus is human. At the early stages of fetal development, the issue is less clear. Most scientists, and just about anyone who's looked at a blastocyst stage embryo under a microscope, lean toward the view that the potential for self-awareness begins further along in embryonic development. Stem cells have yet to differentiate into other kinds of cells; therefore the fetus is not yet on the road to becoming human. There, however, are those who disagree. Many feel that the potential for self-awareness begins at the point of conception, that at every state of gestation the fetus is unequivocally on the road to becoming a human being. From a religious standpoint, many also express the belief that self-awareness begins at conception, when the soul enters or in some way becomes associated with the body. Anything that interferes with the fetus or that supplants, or to some, prevents conception runs counter to the hand of God.

Thus far, nations have taken different positions on therapeutic cloning. South Korea, Great Britain, and other European countries actively support research. Under the Bush Administration, the United States permitted privately funded research but limited the use of federal funds to projects that used stem cell lines from embryos produced in excess by fertility clinics and that would eventually be destroyed. The Obama administration has liberalized this approach and opened the door to a range of federally funded research projects.

Reproductive and therapeutic cloning have brought the natural sciences to a turning point in their development, one so significant that it forces us to reevaluate the limits of science. The natural sciences have always seen the human being in chemical and anatomical terms and have never dealt with the matter of self-awareness. Cloning forces us to confront the esoteric. As scientists and as members of a scientifically enlightened society, cloning challenges us to stretch the bounds of science. It thrusts reflection and our essential humanness to the forefront of scientific thought, wherever this may lead.

45

Abortion

By any measure, abortion is one of the most polarizing issues in American current affairs. It's also an issue that no matter our feelings about abortion need not be as divisive. Common sense allows us to look beyond politics and address the abortion controversy from the standpoint of a practical outcome.

We begin with a brief history of abortion and the politics of abortion.

Traditionally, most societies have held abortion in disfavor, but there have always been abortions. In earlier centuries, women drank poisons, ingested potions, and endured beatings and mutilation in their attempts to end unwanted pregnancy. Done behind closed doors, these efforts often resulted in a woman's death. With the industrial revolution, women began to push for social, political, economic, and reproductive rights. In 1920, women in the United States earned the right to vote. In 1936, a nurse named Margaret Sanger orchestrated a court battle that led to the American Medical Association officially recognizing birth control as part of a doctor's medical practice. In the 1960s, "the pill" hit the market. With the civil rights and women's movement, discrimination based on gender became illegal or socially unacceptable. For the first time in history, women went to college in large numbers and entered the professional workforce. Increased political and economic freedom led to the 1973 Supreme Court *Roe v. Wade* ruling, which held that a woman had the constitutional right to have an abortion—the legal right to end her pregnancy.

Specifically, the Court found that laws against abortion in the United States violated a woman's constitutional right to privacy under the *Due Process Clause* of the *Fourteenth Amendment*. By virtue of a woman's right to privacy, the Court acknowledged that a woman has the right to control what happens to her body, including the reproductive process. In the Court's interpretation, abortions are permissible for any reason a woman may choose until the point at which the fetus becomes "viable," or potentially able to live outside the mother's womb. Furthermore, the court held that abortion after viability must be available to protect a woman's health, which it broadly defined in the companion case of *Doe v. Bolton*.

Effectively, the 1973 Supreme Court rulings granted a woman's right to an abortion; and, over the years, that right has been exercised in abundance. According to the Center for Disease Control, or CDC,[1] in 1970 about 123 thousand women had abortions. That number increased to more than 1.4 million in 1990 and has since declined to an average of about 800 thousand abortions per year. The highest abortion rate takes place in women less than fifteen years old. The lowest rate takes place in women between thirty and thirty-four years old. Younger women also have a higher rate of late-term abortions.

Subsequent to *Roe v. Wade*, the Court has recognized limitations to a woman's right to have an abortion. To date, upwards of forty states have enacted and enforce laws that require parental notification for a minor woman to end an unwanted pregnancy. Viability of the fetus is also a matter of contention. From a legal standpoint, the question is at what time during pregnancy the fetus becomes able to survive outside the womb and, thus, represents a potential citizen for which the state has an obligation to protect. As medical technology has advanced this has become harder to determine. Legislators, and more directly physicians, have placed greater legal and ethical restrictions on late term abortions.

Roe v. Wade has led to an increase in abortions and in the safety of abortions. It has also led to the rise of two opposing political movements. The first labels itself *prochoice*. It's in favor of a woman's right to a legal abortion and directs its political will to fight for and expand the *Roe v. Wade* decision. The second labels itself *prolife*. It seeks to overturn *Roe v. Wade* and fights for the rights of the embryo and fetus and to prevent or restrict a woman's right to a legal abortion. The drawing of political battle lines has also had an unintended consequence—one that perpetuates the debate and that may preclude both sides from achieving victory.

---

1. "Abortion Surveillance—United States 2005," 12.

To understand this, we begin with a look at the reasons doctors perform abortions. An abortion may be conducted to preserve the life or physical wellbeing of the mother. An abortion may be conducted to end a pregnancy that resulted from rape or incest or to prevent the birth of a child with a serious mental, physical, or genetic abnormality. Abortions may be performed to prevent a birth for social or economic reasons, the youth of the woman or the family's financial inability to raise the child. Abortions may also be performed to preserve the mental wellbeing of the patient, and from a mental health standpoint may be considered therapeutic.

On the prochoice side of the political debate, proponents hold that the woman, and not the state, has the right to control what happens during pregnancy. They argue that in Latin America, where abortion is restricted, more than one million women a year seek medical treatment as the result of poorly performed abortions. They also argue in terms of family planning, that a couple and more directly a woman has the responsibility to see to it that every child born is wanted and can be appropriately cared for. In this sense, many prochoice advocates look at abortion as a form of birth control.

Dominant in the prochoice movement is the *National Association for the Repeal of Abortion Laws*, or NARAL, which has become one of the wealthiest and most powerful single-issue lobbying interests in Washington. Also dominant is the *Planned Parenthood Federation of America*. Founded in 1942, and with roots traceable to Margaret Sanger's movement for the medical acceptance of birth control, *Planned Parenthood* in the United States is a nonprofit group that in addition to funding abortion activism provides contraceptives and a wide range of reproductive services. In 2007, Planned Parenthood had revenues in excess of one billion dollars.[2]

On the prolife side of the political debate, the issue centers on the rights of the unborn child. Particularly active in the right-to-life movement are Christian and other religious groups. For many who embrace a religious point of view, human existence begins at the point of conception, and thus abortion is akin to murder. Proponents battle for the legal recognition of the unborn child as a human being with a right to existence at any stage of its development. On the less ideological end of the

---

2. "Planned Parenthood Annual Report 2006-2007," 18.

movement, advocates accept the need for abortion in certain instances but denounce it as a form of birth control.

To further their cause, the prolife moment has taken an even more aggressive stance than that of the prochoice movement. In addition to lobbying efforts intended to restrict access to abortions and to limit the use of federal funds for abortions, groups picket medical and Planned Parenthood offices, and the most radical factions threaten doctors, intimidate patients, and have adopted terrorist tactics and bombed abortion clinics.

As with the environmental movement, the prochoice and prolife movements have followed a familiar developmental pattern. Both factions have evolved into massive, well funded, political enterprises. We have a prochoice industry and a prolife industry devoted to the political and financial maneuverings that pertain to the legality of the abortion issue. This creates a predicament. The fight for or against abortion rights has become the purpose-to-exist for the prochoice and prolife movements—and the massive political and financial machines that these organizations have become. Neither side can succeed in its cause without both going out of business.

And—as the movements incite radical fringes and whip up legislative and public sentiment with messages of rights and ethics, fear and urgency—there are losers: the woman and the unborn child.

When we set aside the politics of abortion, however, a commonsense solution to the issue stands out. With regard to abortion, it, like with the issue of cloning, takes us back to our chapter titled *Self*. At its core, the matter comes down to a simple question: At what point during the development of the fetus do we make the transition from a mass of biological matter to becoming, or to be clearly on the road to becoming, a human being—conscious of our consciousness, potentially able to reflect on our life and on the meaning of our existence? The answer to this question is that no one knows—scientist, politician, or theologian. It may be at the point of conception. Then again, it may be later in the pregnancy. Given this uncertainty, the legal issue of abortion becomes irrelevant. There is a practical solution to the abortion matter. We prevent unwanted pregnancies from happening in the first place.

And there are groups and individuals who have made tremendous strides in this respect, men and women who have devoted their lives to helping teen girls and others gain the maturity to make responsible sexual and reproductive choices. Abortion in the United States has decreased since the 1990s. Still, though, in 2007, 47.8 percent of high school students reported having intercourse at least once and only 61.5 percent reported that they used a condom or some form of birth control. Unwanted pregnancy occurred in 7 percent of women between the ages of 15 and 19 years.[3] We have made strides in ending unwanted pregnancy, but there is an obstacle that prevents us from resolving the matter altogether: the battle that rages between the prolife and the prochoice movements. To resolve the abortion controversy, the prolife and prochoice movements must look beyond the business of the abortion issue—beyond money, power, and political influence. They must put the legal battle aside and unite in the cause to end unwanted pregnancy.

On the prochoice side, proponents have to recognize that abortion isn't just another form of birth control. Granted one-half of fertilized eggs die within ten days of conception and one-fifth of recognized pregnancies end in a miscarriage, but a deliberately induced abortion is another matter. There's more to it than an extension of natural processes. It's a conscious choice to end the life of a potential human being, a future son or daughter. Conversely, prolife advocates—above all those with a religious bent—have to accept the fact that sex has something to do with pregnancy and that contraceptives have something to do with preventing it. We need birth control, and—whether in school or at home, and given the rate of teen pregnancy hopefully both—we need sex education. Abstinence programs may be the ideal way to prevent unwanted pregnancy, but they may not be the most realistic way. The rate of teen sex among women taught to abstain from sex is about the same as the rate among women taught to make their own choice about sex, and teen girls in abstinence-only programs have a higher incidence of unprotected sex. Birth control and sex education need not encourage sex, but they can prevent pregnancy.

No law will stop unwanted pregnancy and abortion. But access to contraception, health education programs for teens that include sex information and that address the social issues of teenage life—including

---

3. Anderson, "Teen Pregnancies at 30-Year Low."

the rationale to abstain from sex—can substantially reduce unwanted pregnancy and abortion.

By devoting their energy to the legal battle for or against abortion, the prolife and prochoice movements have embraced the war as the cause at the expense of the solution the war was in theory launched to provide. The recognition of the nature of the self allows us to wade through politics and emotion and embrace a commonsense approach to resolving the issue of abortion. Both sides in the abortion debate need to pool their resources toward the common goal of ending unwanted pregnancy. Decades more of legal and political wrangling—enriching attorneys, lobbyists, and executives—will not put to rest the abortion issue. Common sense will.

46

UFOs

Our world is crazy, out-of-control, rushing in every possible direction. When it comes to crazy, out-of-control, and rushing in every possible direction, it stands to reason that we'd wrap up the book with a few words on the *New Age* movement and the UFOs, lost continents, Martian civilizations, and everything else it encompasses. From the standpoint of common sense, proponents of the New Age leave much to be desired. But our look at the movement leads us to a concluding point about ourselves.

The New Age movement, which began in the 1960s and is well into its Alzheimer years, has to do with people preaching every imaginable theory and selling every imaginable product. There are New Age politics and New Age religions. There are New Age diets and New Age exercise programs. There are New Age foot magnets, herbal colonics, and nutritional supplements. If it lacks supportability or if it's just plain too out-there to be called anything else, it's New Age. Take the "intelligence" industry, and by intelligence we don't mean what the CIA and Interpol do to fight terrorism. Each year we spend millions of dollars on tests to measure our intelligence and on pills and foods to make us smarter.

To support a claim associated with intelligence, certain assumptions must be true. First, we have to know what intelligence is. Does any scientist have the answer to this question? Is intelligence something we can define or is it a catchall phrase that describes our sense of another's abilities? Perhaps the smartest man or women never thinks about intelligence. It's character and accomplishments that matter.

Second, we have to be able to measure intelligence. Which, if we don't know what intelligence is, can be hard to do. We've all heard the old claim that we use ten percent of our brain. But how do we determine the brain's capacity, and how do we determine how much of it we're using? The people who write IQ tests claim they know how to measure intelligence but only because they define intelligence as a good score on an IQ test. Perhaps, the most intelligent question we can ask is who comes up with this stuff?

People who want to make money, that's who. Most Nobel Prize winners have an average IQ. An IQ test doesn't measure intelligence, its measures our ability to take an IQ test. If you read the label on a dietary supplement claimed by the manufacturer to increase intelligence, you'll usually see a stimulant listed in the ingredients, often caffeine because it's cheap. Intelligence supplements don't increase intelligence, they make us think we're smarter. Which is good, because the only thing they do is make us poorer.

The intelligence industry is a dot on the New Age map. There are so many "alternative paradigms" and they have so little grounding in fact, logic, and common sense that buying into the New Age movement itself may be a measure of intelligence.

The mother of all New Age causes is the UFO; which, for the one person reading who doesn't know, stands for *unidentified flying object*. As anyone who's spent a lot of time outdoors has observed, on occasion there are things visible in the sky that can be hard to explain. We have a lot to learn about atmospheric phenomena and in general about the nature of the universe. But to the UFO people, unexplained phenomena aren't objects of scientific interest. Lights in the sky are spaceships from another planet or dimension. Aliens are here to conduct experiments on us. And you guessed it; these experiments are sexual in nature.

Driven by the success of recent Mars missions, the trendiest New Age cause is that of a lost Martian civilization. Proponents believe that people originated on Mars and long ago faced planetary catastrophe. Martians fled to the earth on spaceships or by spiritual means where they fathered our ancestors and gave rise to humankind. Life may or may not exist on Mars. We don't know. What we do know is that Mars has been mapped and photographed for decades. When we look at the shots of the Martian landscape sent back by our rovers, we see dirt and rocks. The Martian New Agers see footprints and ancient artifacts.

The earth, of course, has its own lost civilizations. When we consider the documented rise of culture, the story is truly remarkable—the advent of art and music, of urbanization and agriculture, of writing, mathematics, and architecture. But there are those who think it needs embellishment. Many are convinced that spiritually and technologically advanced civilizations existed before Ancient Sumer, before the first excavated city, before the first cave paintings and met their demise by sinking into the ocean or through some other cataclysm. Throughout the ages, we've sought the lost wisdom of the ancients.

In all aspects of life, it's imperative to keep an open mind. If archaeological evidence pushes back the date of urbanization from twelve to fourteen thousand years ago, we update our account of civilization. If Martians land on the Whitehouse lawn, we add a chapter to the history books. Neither science nor philosophy has all the answers, and new information continually emerges. But we must filter what we read and hear through our common sense.

The proponents of the New Age, however, are good at encouraging us to avoid this step. In particular, they're masters at manipulating the technique of the *Argument*.

Ancient buildings have high ceilings and huge doorways; therefore, the earth was inhabited by giants. The average weight of the stone blocks used to build the pyramids at Giza is 2.5 tons. The Egyptians couldn't move blocks this big; therefore, the pyramids were built by extraterrestrials. Archeology tells us that the New World was inhabited before the Bering Strait land bridge formed ten thousand years ago; therefore, North and South America were populated by the lost civilization of Atlantis.

Each of these arguments is presented in textbook form. But, as we learned in the first chapter, for an argument to be reasonable, the premises must be true and they must logically lead to the conclusions. We make buildings with high ceilings and big doors today; and, basketball players and steroid use aside, there are no giants. People can move stone blocks much bigger than 2.5 tons, about the weight of an SUV, with levers and pulleys, and the ancient Egyptians had levers and pulleys. Archaeological evidence says that people may have inhabited the New World as early as thirty thousand years ago. Does this mean they came from Atlantis?

The physical evidence cited in support of a New Age theory is invariably something we convince ourselves to see or something we can never see. The Dead Sea Scrolls and the spiritual healer Edgar Cayce talk of ancient civilizations, but the ruins of these civilizations are on the bottom of the ocean. The Bible and the prophet Nostradamus predict global catastrophe and the end of the world as we know it, and the environmentalists warn of global warming and ecological collapse. In 1945, a UFO crashed at Roswell New Mexico, but the government covered it up.

The point of this chapter isn't to be critical of the New Age movement and its participants; we each have our own path in life. Neither is it to end the book with humor. The New Age leads us to a question about ourselves. Each year, millions buy New Age books. Each night, millions tune into "alternative" radio to listen to callers share stories about prophecy, the end-times, and alien abduction. What is it about modern society that makes the New Age appealing? Do we have so little hope in life that the best we can do is dream of rediscovering the wisdom of the ancients or of being whisked to another planet or dimension on a UFO?

Bibliography

"Abortion Surveillance—United States 2005." *Center for Disease Control Morbidity and Mortality Weekly Report Surveillance Summaries* 57 SS13 (November 28, 2008). Online: http://www.cdc.gov.

"Analysis of the Lieberman-Warner Climate Security Act (S.2191) using the National Energy Modeling System (NEMS/ACCF/NAM)." Washington, D.C.: The National Association of Manufacturers and the American Council for Capital Formation, 2008. Online: http://www.accf.org./media/dynamic/1/media_190.pdf.

Anderson, Lisa. "Teen Pregnancies at 30-Year Low." *Chicago Tribune* (June 29, 2008) 3. Online: Chicagotribune.com.

Andrews, Charles. *From Capitalism to Equality: An Inquiry into the Laws of Economic Change*. Oakland, CA: Needle Press, 2000.

Armstrong, Karen. *Holy War: The Crusades and Their Impact on Today's World*. New York: Anchor Books, 2001.

Beck, Glenn. *An Inconvenient Book: Real Solutions to the World's Biggest Problems*. New York: Simon & Schuster, 2007.

Bedogne, Vincent F. *Blueprint for Reconstruction: The Rebuilding of the Earth's Urban and Ecological Infrastructure*. Threshold to Meaning. Eugene, OR: Wipf and Stock, 2009.

———. *Economics of Fulfillment: The Obsolescence of Socialism and Capitalism and an Economic Philosophy for the Future*. Threshold to Meaning. Eugene, OR: Wipf and Stock, 2009.

———. *Evolution of Consciousness: The Philosophy of Pierre Teilhard de Chardin and the Evolutionary Transformation Unfolding Within Us*. Threshold to Meaning. Eugene, OR: Wipf and Stock, 2008.

"A Better Way Than Cap and Trade." *Washington Post* (June 26, 2008) A19. Online: http://www.washingtonpost.com/wp-dyn.

Brockway, George. *The End of Economic Man: An Introduction to Humanistic Economics*. New York: Norton, 2001.

Brown, E. and R. B. Firestone. *Table of Radioactive Isotopes*. New York: Wiley Interscience, 1986.

Brown, Theodore L., et al. *Chemistry: The Central Science, 6th ed*. New Jersey: Prentice Hall, 1994.

Clausewitz, Carl Von. *On War*. Translated by J. J. Graham. New York: Penguin, 1982.

Cline, Mary. "The Health Care System I Want is in France." *ABC News* (April 15, 2009). Online: http://abcnews.go.com/Health/story?id=4647483&page=1.

Corbett, J. O. "The Radiation Dose From Coal Burning: A Review of Pathways and Data." *Radiation Protection Dosimetry* 4 (January 1983) 5–19.

Danhof, Justin. "Why Cap and Trade Could Backfire." *Christian Science Monitor* (July 16, 2008). Online: http://www.globalpolicy.org/component/content/article/216-global-taxes/45869.html.

Dutton, Paul V. "Health Care in France and the United States: Learning from Each Other." Washington, D.C.: Brookings Institution, 2008. Online: http://www.brookings.edu/~/media/Files/rc/articles/2002/07france_dutton/dutton.pdf.

"Exposure of the Population in the United States and Canada from Natural Background Radiation, Report 94." Bethesda, MD: National Council on Radiation Protection, 1987.

Friedman, Thomas L. *From Beirut to Jerusalem.* New York: Anchor Books, 1995.

Gabbard, Alex. "Coal Combustion: Nuclear Resource or Danger." Oakridge, TN: Oak Ridge National Laboratory, 1993. Online: http://www.pushback.com or http://www.ornl.gov/info/ornlreview/rev26-34/text/colmain.html.

Giere, Ronald N. *Understanding Scientific Reasoning, 3rd ed.* Orlando: Holt, Rinehard, and Winston, 1991.

Hannity, Shawn. *Deliver us from Evil: Defeating Terrorism, Despotism, and Liberalism.* New York: Harper Collins, 2004.

"The Heat is On: Gore's 'Carbon Offsets' paid to a firm he owns." *WorldNetDaily.com* (March 2, 2007). Online: http://www.worldnetdaily.com/news/article.asp?ARTICLE_ID=54528.

Hourani, Albert. *A History of the Arab Peoples.* New York: Warner Books, 1991.

"How it Works: Cap-and-Trade Systems." *Catalyst* 4 (Spring 2005) 18–19. Online: http://www.ucsusa.org/assets/documents/catalyst/Catalyst-Spring-2005.pdf.

Hurley, Patrick J. *A Concise Introduction to Logic, 5th ed.* Belmont: Wadsworth Publishing Company, 1994.

"Hybrid Hype: The Dollars & Sense of Hybrids." *Consumer Reports* 71 (April 2006) 18–22.

Isikoff, Michael and David Corn. *Hubris: The Inside Story of Spin, Scandal, and the Selling of the Iraq War.* New York: Three Rivers Press, 2007.

Jantsch, Eric. *The Self-Organizing Universe: Scientific and Human Implications of the Emerging Paradigm of Evolution.* New York: Pergamon Press 1980.

Jones, Steven, et al. *The Cambridge Encyclopedia of Human Evolution.* Cambridge: Cambridge University Press, 1992.

Judkins, R. R., et al. "The Dilemma of Fossil Fuel Use and Global Climate Change." *Energy & Fuels* 7 (1993) 14–22. Online: http://pubs.acs.org/doi/abs/10.1021/ef00037a004.

Kanath, Rajani. *Against Economics: Rethinking Political Economy.* Farnham, UK: Ashgate, 1997.

Keynes, John Maynard. *The Economic Consequences of Peace.* New York: Harcourt, Brace and How, 1920.

———. *The General Theory of Employment, Interest and Money.* New York: Harcourt Brace, 1935.

———. *A Treatise on Money.* New York: AMS Press, 1976.

Krane, Kenneth. *Modern Physics, 2nd ed.* New York: John Wiley & Sons, 1996.

Land, Kenneth C, Project Coordinator. "The 2008 Foundation for Child Development Child and Youth Well-Being Index (CWI) Report." Durham, NC: Duke University, 2008. Online: http://www.fcd-us.org/usr_doc/2008AnnualRelease.pdf.

Lartigue, Casey, and Ryan Balis. "The Lieberman-Warner Cap and Trade Bill: Quick Summary and Analysis." *National Policy Analysis* 570 (June 2008). Online: http://www.nationalcenter.org.

"Levels and Trends in Contraceptive Use as Assessed in 2002." New York: United Nations Department of Economic and Social Affairs Population Division, 2006.

Levin, Mark R. *Liberty and Tyranny: A Conservative Manifesto.* New York: Simon & Schuster, 2009.

Lewis, Bernard. *The Middle East: A Brief History of the Last 2,000 Years.* New York: Touchstone, 1997.

Lieberman, Ben. "Five Myths About the Lieberman-Warner Global Warming Legislation." Washington, D.C.: Heritage Foundation, 2008. Online: http://www.heritage.org/Press/Commentary/ed060308a.cfm.

Lieberman, Joseph. "Fighting Global Warming the Right Way." *Hartford Courant* (October 22, 2007). Online: http://lieberman.senate.gov/newsroom/release.cfm?id=285748.

Lomborg, Bjorn. "The Climate Industrial Complex." *Wall Street Journal* (May 22 1999). Online: http://online.wsj.com/article/SB124286145192740987.html.

———. *Cool It: The Skeptical Environmentalist's Guide to Global Warming.* New York: Knopf, 2007.

Malthus, Thomas R. *Definitions in Political Economy.* New York: A. M. Kelly, 1963.

———. *An Essay on the Principle of Population.* Homewood: R. D. Irwin, 1963.

"Many Americans Think an AIDS/HIV Vaccine Already Exists." *National Institute of Allergy and Infectious Diseases (NIAID) News* (May 15, 2003). Online: http://www.nih.gov/news/pr/may2003/niaid-15.htm.

Marx, Karl. *Capital, A Critique of Political Economy.* New York: The Modern Library, 1906.

———. *The Communist Manifesto.* Chicago: H. Regency Co., 1954.

———. *A Contribution to the Critique of Political Economy.* New York: International Publishers: 1970.

Michaels, Patrick. "Cato Scholar Comments on Warner-Lieberman Climate Security Act." Washington, D.C.: Cato Institute, 2008. Online: http://www.cato.org/pressroom.php?display=ncomments&id=34.

"The Mideast: A Century of Conflict." *National Public Radio* (September 2002). Online: http://www.npr.org/news/specials/mideast/history.

Mill, John Stuart. *Principles of Political Economy.* New York: McGraw Hill, 1973.

Morris, David. "Al Gore's Carbon Solution Won't Stop Climate Change." *Environment* (March 12, 2007). Online: http://www.alternet.org/environment/49025.

Muller, Richard A. *Physics for Future Presidents: The Science Behind the Headlines.* New York: W. W. Norton, 2008.

Murray, Brian C. and Martin T. Ross. "The Lieberman-Warner America's Climate Security Act: A Preliminary Assessment of Potential Economic Impacts." Durham, NC: Duke University Nicholas Institute for Environmental Policy Solutions, 2007. Online: http://www.nicholas.duke.edu/institute/econsummary.pdf.

Norris, Robert E. and L. Lloyd Haring. *Political Geography.* Columbus, OH: Charles E. Merril and Company, 1980.

"Overwhelming Majority of Americans Oppose Lieberman-Warner Global Warming Proposal, New Poll Suggests." Washington, D.C.: National Center for Public Policy Research, 2008. Online: http://www.nationalcenter.org.

Paltsev, Sergey, et al. "Assessment of U.S. Cap-and-Trade Proposals." Cambridge, MA: MIT Joint Program on the Science and Policy of Global Change, 2007. Online: http://w3.mit.edu/globalchange.

Peterson, Willis. *Principles of Economics, 4th ed.* Homewood: Richard D. Irwin, 1980.

Planned Parenthood Federation of American Annual Report 2007 – 2008. New York: Planned Parenthood, 2009.

"Public Radiation Exposure From Nuclear Power Generation in the U.S., Report 92." Bethesda, MD: National Council on Radiation Protection, 1987.

Putterman, Louis. *Dollars and Change: Economics in Context*. New Haven: Yale University Press, 2001.

"Radiation Exposure of the U.S. Population from Consumer Products and Miscellaneous Sources, Report 95." Bethesda, MD: National Council on Radiation Protection, 1987.

Resnick, Robert and David Halliday. *Basic Concepts in Relativity Theory and Early Quantum Mechanics*. Englewood Cliffs: Prentice Hall, 1991.

Ricardo, David. *Principles of Political Economy and Taxation*. New York: Dutton, 1973.

Schneider, Herbert Wallace. *Adam Smith's Moral and Political Philosophy*. New York: Harper and Row, 1948.

Schumpeter, Joseph. *Business Cycles: A Theoretical, Historical, and Statistical Analysis of the Capitalistic Process*. New York: McGraw Hill, 1964.

Serway, Raymond A. *Principles of Physics*. New York: Harcourt, 1994.

Smith, Adam. *An Inquiry Into The Nature and Causes of the Wealth of Nations*. New York: P. F. Collier and Son, 1910.

———. *The Theory of Moral Sentiments*. Indianapolis: Liberty Classics, 1976.

Stikker, Allerd. *The Transformation Factor: Towards an Ecological Consciousness*. Rockport: Element Books, 1993.

"Study: Cap-and-Trade Won't Slow Economy." *Environmental Protection* (April 2008). Online: http://www.eponline.com/articles/61412.

"Teen Birth Rate Rises in US Reversing a 14-Year Decline." *Washingtonpost.com* (December 2007). Online: http://www.washingtonpost.com/wp-dyn/content.

Teilhard de Chardin, Pierre. *The Phenomenon of Man*. Translated by Bernard Wall. New York: Harper and Row, 1959.

Torrey, W. "Coal Ash Utilization: Fly Ash, Bottom Ash and Slag." *Pollution Technology Review* 48 (1978) 136.

Townsend, Mark and Jason Burke. "The Earth Will Expire by 2050." *Observer of London* (July 7, 2002). Online: http://www.commondreams.org/headlines02/0707-02.htm.

"Transitions in World Population." *Population Reference Bureau Population Bulletin 59* (2004). Online: http://www/prb.org.

"US Health Care Expensive, Inefficient." *MSNBC.com* (May 2007). Online: http://msnbc.msn.com.

"US Life Expectance Lags Behind Other Countries." *CNN.com* (August 2007). Online: http://cnn.com.

William W. Beach, et al. "The Economic Costs of the Lieberman-Warner Climate Change Legislation." Washington, D.C.: Heritage Center for Data Analysis, 2008. Online: http://www.heritage.org/Research/EnergyandEnvironment/upload/cda_0802.pdf.

"World Population Prospects The 2002 Revision." New York: United Nations Department of Economic and Social Affairs Population Division, 2003. Online: http://www.un.org/esa/population/publications/wpp2002/WPP2002-HIGHLIGHTSrev1.PDF.

www.ingramcontent.com/pod-product-compliance
Lightning Source LLC
Chambersburg PA
CBHW070248230426
43664CB00014B/2447